Charles Darwin, the Copley Medal, and the Rise of Naturalism 1861–1864

"Reacting to the Past" Series

Charles Darwin, the Copley Medal, and the Rise of Naturalism 1861–1864

Marsha Driscoll, Elizabeth E. Dunn, Dann Siems, and B. Kamran Swanson

in consulation with

Frederick H. Burkhardt

W. W. Norton & Company
New York London

W. W. Norton & Company has been independent since its founding in 1923, when William Warder Norton and Mary D. Herter Norton first published lectures delivered at the People's Institute, the adult education division of New York City's Cooper Union. The firm soon expanded its program beyond the Institute, publishing books by celebrated academics from America and abroad. By mid-century, the two major pillars of Norton's publishing program—trade books and college texts—were firmly established. In the 1950s, the Norton family transferred control of the company to its employees, and today—with a staff of four hundred and a comparable number of trade, college, and professional titles published each year—W. W. Norton & Company stands as the largest and oldest publishing house owned wholly by its employees.

Production Manager: Ashley Horna

Library of Congress Cataloging-in-Publication Data

Driscoll, Marsha, author.
    Charles Darwin, the Copley medal, and the rise of natualism, 1862–1864 / Marsha Driscoll, Elizabeth E. Dunn, Dann Siems, B. Kamran Swanson ; in consultation with Frederick H. Burkhardt.
        pages cm.—(Reacting to the past)
    Reprint of: Boston : Longman, ?2010.
    Includes bibliographical references.
    ISBN 978-0-393-93726-8 (paperback)
    1. Natural History—England—History—19th century.   2. Darwin, Charles, 1809–1882.   3. England—Intellectual life—19th century.   4. Royal Society (Great Britain).   5. Science—Philosophy.   6. Natural Selection.   I. Title.
    QH21.G7D75 2013
    576.8'2—dc23
                                                                            2013042599

W. W. Norton & Company, Inc., 500 Fifth Avenue, New York, NY 10110
wwnorton.com

W. W. Norton & Company Ltd., Castle House, 75/76 Wells Street, London W1T 3QT

1 2 3 4 5 6 7 8 9 0

INFORMATION AND CREDITS

ABOUT THE GAME

Charles Darwin, the Copley Medal, and the Rise of Naturalism is part of the "Reacting to the Past" Series, a pedagogical initiative offered under the auspices of Barnard College. Faculty or college administrators interested in more information on the "Reacting to the Past" Series should send an e-mail to reacting@barnard.edu.

This packet constitutes one component of the *Charles Darwin, the Copley Medal, and the Rise of Naturalism* game. Other materials include access to a web site that provides additional information relevant to the game, as well as individual student roles for the game [the full version of the *Charles Darwin, the Copley Medal, and the Rise of Naturalism* includes 21 separate roles, for a total (maximum) class of 25 students]. These roles are not, and cannot be, publicly distributed. Other teaching materials include an Instructor's Manual for the game, and the *Reacting Pedagogy Manual*. Janet Brown's (2007) *Darwin's "Origin of Species": A Biography* is highly recommended as background material.

The instructor's manual and roles can be obtained through the Pearson Education Instructor Resource Center or directly through the "Reacting to the Past" Program office at Barnard College (see http://www.barnard.edu/reacting for additional information).

ACKNOWLEDGMENTS

Charles Darwin, the Copley Medal, and the Rise of Naturalism has benefited enormously from the suggestions, questions, and research of many faculty and students. Special thanks go to Keith Kester of Colorado College for his willingness to test play the game with the students in his Science and Religion course. The authors also wish to thank Douglas Allchin from the University of Minnesota for his unflagging support and many useful comments.

CONTENTS

THE GAME   **1**

Introduction: Welcome to Victorian England   **3**

The Natural Theologians   5

The Naturalists   6

The Social Reformers   7

Basic Principles   **9**

The Play of the Game   **10**

Game Setting   10

Royal Society Meetings   11

Your Role in the Royal Society   11

The Copley Medal   13

Special Roles in the Council of the Royal Society   13

Special Rules   **15**

Retention of Seat on the Council   15

Disqualification for reading aloud   15

Role of Gamemaster, Contact with Instructor   15

Student-Initiated Rule Modifications   16

A Word on The Use of Props   16

The Main Factions   **17**

A-Men (opponents of Darwin)   17

X-Men (supporters of Darwin)   20

Brief Sketches of Game Characters   **22**

Historical and Composite Factional Characters   22

Indeterminate Roles   23

Proceedings of the Royal Society   **26**

Podium Rule   26

Copley Nominations   26

Prelude to the Game   26

Detailed Agenda (Session by session)   27

Summary table of agenda and assignments: 29

Protocol and Parliamentary Procedure   31

Specific Written and Oral Assignments   **32**

Playing a Natural Philosopher or Man of Science in   **33**
Victorian England

Introduction to the Philosophical Controversy   34

The Historical Context: Things You Should Know   44

APPENDICES   **53**

Appendix A. Darwin, *On the Origin of Species* (1859)   **55**

Appendix B. Primary Source Documents   **117**

Samuel Wilberforce, Review of *On the Origin of Species*   118

John Lubbock, "Tact"   143

Francis Bacon's *Novum Organum*, 1620 (excerpts)   148

William Paley, *Natural Theology; or, Evidences*   155
of the Existence and Attributes of the Deity

Charles Kingsley, "A Nation's Grief for a Nation's Loss"   160

Sample Prayers from *The Book of Common Prayer*   162

Song Lyrics: "God Save the Queen"   164

Song Lyrics: "All Things Bright and Beautiful"   165

Appendix C. Additional Sources   **166**

The Game

INTRODUCTION: WELCOME TO VICTORIAN ENGLAND

It is early January, 1862. You are a member of the British elite, most likely either part of the gentry (property owning individuals who are independently wealthy) or the growing professional class. You have just traveled to London on one of the modern railway passenger cars from your home in one of the counties and you contemplate that the journey is certainly much easier than it was in your youth, when you would have endured at least two days of rutted toll roads, poorly sprung carriages, and the threat of highwaymen.

As you step off the railway car and onto the platform at Waterloo Station, you realize with amazement that, although the noisy steam engine has stopped running, you are still overwhelmed by the noise around you. London is teeming with people and the dark oppressive air is not the result of the train you have been riding; rather, it is the permanent haze that has been thickening around London for many years now. Since it is midday, you are not yet privileged to see the amazing effect of the new gas lamps that have apparently turned night into day in the heart of London, but you wonder if one ever actually sees the sun through this yellow-orange fog.

Deciding to walk rather than take a hackney cab through the crowd, you travel quickly across several city streets, pay a toll at the Waterloo Bridge (hoping you don't witness any suicides as you cross the Thames), and finally reach the busy Strand, home to a variety of shops, several publishing houses, and newspaper offices. You make a short detour to pass the Adelphi Theatre that is currently featuring the Anglo-Irish dramatist, Dion Boucicault. You hope to be able to see his adaptation of one of Charles Dickens's current works while you are in town.

You reach the Royal College of Surgeons and you dash up the steps to see which lectures are scheduled over the next few days. You are in luck! A student of Professor Richard Owen, the noted anatomist/paleontologist, will deliver an address on the comparative anatomy of Chinese Shanghai fowl and Sussex game fowl. Since this seems to be an ornithological session, you are hopeful that the esteemed professor, himself, may mention his latest fossil finds, which you have heard are of some type of flying dinosaur. The study of birds is a favorite of many men of science. The current president of the Royal Society, General Sir Edward Sabine, ventured away from his astronomy and geophysics to make a taxonomic study of birds of North America. And, recently, Charles Darwin used the many varieties of specially bred pigeons to support his case for common ancestry of various species. You make note of the time for the lecture tomorrow afternoon.

You also see a notice concerning the exhibit at the British Museum of the sculptures from the Mausoleum at Halikarnassos in Turkey. You're slightly irritated since they have been publicizing the same exhibit for the past two years and you weren't all that interested the first time you saw the sculptures. You would much rather see the rapidly expanding natural history collection, but the last time you visited it, the rooms were so crowded with skeletons and prepared specimens that you could hardly move about.

You decide not to spend your afternoon walking among the Egyptian and Asian artifacts that crowd the museum and instead you hail a cab and head toward Burlington House in Piccadilly. Burlington House is the site of the Royal Society, that venerable organization of men of philosophy and letters who are devoted to the study of science. (Although the word "scientist" is occasionally used by some in an effort to draw attention to the particular emphasis of such scholars, you find the term slightly derogatory, or at least somewhat lacking in gentility.) Technically, the Royal Society is considered a gentlemen's club, but, unlike White's or Boodle's, it is much more than a place to eat, drink, and gamble. Even the politically-charged Brooks, where many Whigs meet to discuss their reformist ideas, cannot compare to the serious nature of the scientific discussions at the Royal Society.

You are proud of Burlington House. It represents the progressive nature of Her Majesty Queen Victoria's government and especially of her late husband Prince Albert. If you were to continue west on Piccadilly Street, you would come to the Crystal Palace, the site of the 1851 Great Exhibition, which was built at the insistence of Prince Albert. Proceeds from the Great Exhibition supported the Royal Society, as well as the Natural History collection of the British Museum. In fact, a grant of £1000 from the government in 1850 allowed the Royal Society to provide funds to its members to do research and to buy equipment. In the decade since then, the Royal Society has continued to obtain research funding from the government and private sources. You and other men of science are hopeful of receiving some of that money to support your various expeditions and researches.

It is a relief to leave the teeming streets of the City, with its beggars and thieves. The overcrowding and rapid expansion of London's poor seems to support Thomas Malthus's ideas on poverty and you remind yourself of his words: "Population, when unchecked, increases in a geometrical ratio. Subsistence increases only in an arithmetical ratio. A slight acquaintance with numbers will show the immensity of the first power in comparison of the second" (*An Essay on the Principle of Population,* London, 1798). You suppose that if something isn't done soon, the entire population of England may starve.

As you put yourself farther away from the River Thames, you wonder if perhaps another solution to the population problem may present itself naturally. The disgusting smell that comes from the polluted water is considered by many to be the source of the last outbreak of cholera in the city a short six years earlier. Many of your scientific colleagues support the idea that disease results from *pythogenesis*, its spontaneous occurrence from filth. You have a few questions about that idea, since you live near many farms that contain plenty of dirt and excrement, but on which the families are quite healthy. Perhaps you will bring up the matter for discussion at one of your scientific clubs while you are in town.

As you enter Burlington House, you realize that there is an excitement, a tension, in the air. Members are grouped together speaking in agitated voices, congratulating each other with handshakes and slaps on the back, and then quickly looking around as if to see who might be noticing them. There appear to be at least two separate groups, both claiming some sort of victory. Your curiosity heightened, you find one of your friends to ask what is happening.

"It's the inexhaustible aftermath of the Oxford Debate," he exclaims. "A year and a half later, both sides are still declaring victory. The death of the Prince has rekindled the debate and raised the stakes even higher. Now, there's rumor that Falconer and Huxley are going to nominate Darwin for the Copley Medal this spring."

You look around and notice "Soapy Sam" Wilberforce accepting congratulations for his "final victory" over heresy, while Thomas Huxley celebrates the final demise of "prodigious ignorance and thoroughly unscientific habit of mind."

As amazed as you are to realize that both sides of the argument are still declaring victory, you are even more astonished at the degree of hostility that seems to be present so many months after the debate. These are men of science; they have held many arguments regarding the details of their ideas since the inception of the Royal Society in 1622; surely things are not as vicious as they seem on the surface.

THE NATURAL THEOLOGIANS

You first stop near the group listening to Bishop Samuel Wilberforce, who agrees that William Paley's argument from design still reigns supreme. Clearly, the recent death of Prince Albert must somehow be a part of God's larger, mysterious plan. The Church of England has a long tradition of supporting scientific study as a way to more clearly understand the mind of God and Wilberforce still confidently denies that Darwin's publication *On the Origin of Species* has changed that truth. Darwin's evidence flies in the face of religious doctrine and must, therefore, be incomplete and inaccurate. It is simply impossible that the order of the universe is the result of caprice and it is hard to believe that human beings share common ancestors with the *orangutans* that recently exhibited at the London Zoological Gardens. Richard Owen, the renowned anatomist, is smiling at the Bishop and occasionally amending Wilberforce's science. Wilberforce refers occasionally to the Great Chain of Being in which every creature, animate or even inanimate, from angels to rocks, has a place determined by God's plan.

"No matter how many examples of finch beaks Charles Darwin examines, he cannot explain the exquisite intricacy of design found in the eye," intones the smooth speaking bishop. "Remember Paley's words: "Observe a new-born child first lifting up its eyelids. What does the opening of the curtain discover? The anterior part of two pellucid globes, which, when they come to be examined, are found to be constructed upon strict optical principles; the self-same principles upon which we ourselves construct optical instruments" (*Natural Theology*, p. 6). Does Mr. Darwin believe that these optical principles can truly have been the result of accident? It is absolutely self-evident that the eye is the result of intelligent, purposeful design, not the result of accident and luck! And to say that all the various creatures in God's kingdom came from some common ancestor is ludicrous. How can one possibly claim that a horse is cousin to a fish? The man's so-called science is a delightful bit of fiction and fancy!"

Owen looks a bit uncomfortable with Wilberforce's last comment and makes the comment that of course the horse and the fish actually do share a fair degree of similarity. "Only remember my recent work on the vertebral archetype," he reminds the listeners. "Fins and limbs are essentially the same basic type of anatomy, but that does not imply any common ancestor. Instead, they both reflect the one perfect idea of a vertebrate which exists in the mind of God and they, in their imperfection, inhabit this earthly sphere." There are nods and murmurs of approval as Wilberforce interrupts, "Thus the many species provide ample evidence of the fall of man and the infinite compassion of God." You listen for a while longer and then move on to another group.

THE NATURALISTS

Thomas Huxley is not waiting for others to congratulate him; he is congratulating himself. He is already earning for himself the name of "Darwin's Bulldog," by his insistence that Darwin's theory is unassailable and will pull down the walls of the established (and to his mind corrupt) Church. You know the bias of Huxley's anti-religious sentiments, but you also know that he is a sound scientist. Huxley has begun advocating a religious position which he calls "agnosticism," while insisting that Darwin's argument in favor of natural selection as the explanation for the existence of species is "simple, obvious, and brilliant." You are especially interested to see the more genteel and refined Joseph Hooker and the wealthy young patrician John Lubbock supporting Huxley's point of view.

You listen to some grumbling among a number of your physician friends who may not agree with Huxley, but who certainly don't like Wilberforce. The recent, if brief, recovery of Albert, Queen Victoria's Prince Consort, from the ravages of typhoid was marked by a thanksgiving church service from which the medical community was deliberately excluded. The churchmen eagerly took full credit for the Prince Consort's recovery, claiming that the National Day of Prayer that they instituted caused the change in the Prince's health. You wonder what they are saying now that the Prince has died and left Her Majesty so seemingly despondent. Prince Albert himself was a great supporter of scientific discovery and invention; you cannot imagine that he would have approved of any religious condemnation of medical science. Across the room you hear the physicist, Professor John Tyndall, exclaiming that it is time someone put this entire question of the "efficacy of prayer" to the test.

Someone you do not recognize shouts back at Tyndall: "You sound like those heretics in *Essays and Reviews* who read the Bible as if it is nothing more important than Catullus' poetry. You can't go around claiming that paleontology will explain Genesis."

"Hah!" Huxley rejoins, "Better that than expecting Genesis to explain anything about paleontology or comparative anatomy!"

THE SOCIAL REFORMERS

As you glance around the ornate rooms of Burlington House, you realize that many of those present seem more confused than hostile. Perhaps a few members have not yet read Charles Darwin's latest publication, although by now certainly most have, and nearly everyone has read his highly enjoyable and unobjectionable *Voyage of the Beagle.* Darwin himself is not to be seen. He seldom leaves his home of Down House in Kent to travel to London these days, and he seems content to conduct his research and leave the controversy and politicking to others of more robust health and combative temperament. You wonder if perhaps his absence reflects political astuteness more than a reclusive personality, since Darwin has always struck you as being a rather clever gentleman.

Many of those present are more interested in issues other than Darwin's theory. They are concerned with social issues, economics, politics, women's rights, social status, education, not to mention advancing their personal prospects and social standing. John Stuart Mill and the radical women with whom he associates have begun talking about allowing women the right to vote when, barely four years ago, women who were legally separated from their husbands gained the right to control their own earnings! You have heard that Florence Nightingale was recently elected to the Statistical Society for her ingenious development of polar diagrams. You wonder if she might even be nominated for membership in the Royal Society for the amazing work she did on conditions in military hospitals in the Crimea a few years ago. You suspect that your physician friends might find themselves in agreement with Bishop Wilberforce if the question arises about allowing women into this club.

On the other hand, nearly everyone in the Royal Society approves of the use of science to deal with issues of poverty and sanitation in the London slums. You just aren't all in agreement about what science has actually proven about the origin of these problems or the nature of appropriate solutions. Do the poor suffer because it is part of God's plan? Do they suffer because they are simply inferior examples of the human species and will die because of the process of natural selection? In either case, what is a responsible Englishman supposed to do about it? When it comes to social issues, alliances cross over the divisions created by Darwin's new approach to natural science. You will need to familiarize yourself with these agendas in order to navigate the many swift and shifting currents of the Royal Society's organizational waters.

As you try to decide which group to join, the major-domo asks if you wish to be seated for dinner, and you allow yourself to be moved into the tasteful dining room with the hope of finding friends with whom to discuss these portentous current events. You are aware that you may be watching a pivotal moment of history, with the future of science, the future of the Church of England, and perhaps even the future of the empire itself being influenced by these fellows with whom you now share a glass of port and a cigar. Your thoughts and words may make it into the minutes of the Royal Society and perhaps into the history books. You rub your hands in anticipation of the intellectual challenges ahead of you. It is time to join the fray!

BASIC PRINCIPLES

The purpose of this "Reacting" game is to engage students with ideas central to one of the most profound and culturally transformative intellectual conflicts of the nineteenth century. Throughout Great Britain, her empire, and Western Europe, prevailing opinion held that both natural and social order reflected a Divine Plan. From this perspective, all strife, disease, and poverty arose as a direct consequence of human separation from that plan because of original sin. In direct contrast, a wealth of new scientific evidence from natural history began to make such theologically-inspired beliefs increasingly untenable. Charles Darwin not only collected much of this evidence, but also recognized its implications.

The ultimate victory goal in this game is to control whether or not the Royal Society awards Charles Darwin its prestigious Copley Medal to recognize his empirical and theoretical contributions to natural history, particularly his theory of evolution by natural selection as presented in On the *Origin of Species*. Awarding the Copley Medal to Darwin can be interpreted as symbolic of a scientific endorsement of Darwin's naturalistic views and thus would constitute a direct affront to theological and cultural orthodoxy. Additionally, students will direct their efforts towards achieving specific victory goals, i.e., passing resolutions on related matters, such as promoting scientific study of the efficacy of prayer; supporting or condemning *Essays and Reviews* and/or the "Students' Declaration of Fellows of the Natural and Physical Sciences;" supporting elimination of poverty; calling for a complete end to slavery; and advocating for the election of women to the Royal Society.

Effective participation in this game requires (1) a clear understanding of the premises and arguments put forward by Darwin (1859) in *On the Origin of Species*, and (2) a clear understanding of the premises and arguments of *Natural Theology* as argued by William Paley (1802) and the authors of the Bridgewater Treatises (1833–1840). In addition, players will need to develop an awareness of how these contrasting worldviews interacted with the major social, political, and economic issues of the time.

THE PLAY OF THE GAME

The action takes place in meetings of the Royal Society in London, England, during the years 1862–1864. In 1859, Charles Darwin's long awaited treatise *On the Origin of Species* finally appeared in published form. Because Darwin's theory of evolution by natural selection presented a plausible and wholly naturalistic explanation of order in nature, it presented a direct challenge to widely accepted theological views. For the next five years, there followed a vigorous, complex debate within the scientific community that exemplified the struggle involved in a "paradigm shift" away from Natural Theology toward materialistic naturalism. Thomas Henry Huxley and Bishop Samuel Wilberforce presented arguments for and against the theory in a dramatic and highly public encounter at the annual meeting of the British Association for the Advancement of Science at Oxford in 1860. The game opens a year and half later in 1862, shortly after the death of Prince Albert. This game contracts and distills the ensuing debate and culminates with the 1864 vote for the Copley Medal.

Darwin's theory of evolution instigated sweeping changes in nearly every field of human understanding, including the nature of science itself. The debate over whether or not the Council of the Royal Society should formally recognize Darwin's transformative contributions became imbued with concerns far beyond the merit of his theory and expanded to disagreements about the professionalization of science, the supposed superiority of the white (and especially the English) race, Britain's appropriate role in the ongoing Civil War in the United States, the role of theology in scientific study, the basis of economic differences between the classes, and the nature and rights of women.

GAME SETTING

This game is set in London, England, beginning in January of 1862, three weeks after the death of the Prince Albert, Consort of Queen Victoria, and it continues through November of 1864. Prior to beginning the game, students may watch a video segment enacting the proceedings of the 1860 Oxford meeting of the British Association for the Advancement of Science. At that meeting, which closely followed the publication of Darwin's *On the Origin of Species*, Wilberforce and Huxley held a vigorous, highly contentious debate regarding Darwin's book. The debate set the stage for several simultaneous shifts: from Natural Theology to natural science, from amateur scholarship to professional academic disciplines, from accepting on faith a preordained fixed and orderly universe to contemplating one which is mutable and perhaps without any pre-ordained order at all! Each game session takes place in Burlington House, the site of the meetings of that highly selective and intellectual gentleman's club, the Royal Society.

ROYAL SOCIETY MEETINGS

Two types of Royal Society (RS) meetings will be held during the game. General meetings of the full society provide a forum for scientific papers and the pursuit of overall educational objectives in the game. These meetings would have been attended by many of the members, not just the Council, and will be presided over by General Sir Edward Sabine. Faculty and preceptors may act as additional members of the RS during general meetings. Question and answer sessions will follow paper presentations.

General Sabine will also call regular council meetings. In the game, all players are members of the Council, eligible to attend council sessions and vote on resolutions. Here, substantive debate over a variety of issues may take place. Here, too, is where decisions will be made about the merits of scientific research and matters of public policy, as well as the ultimate decision regarding the Copley Medal. Resolutions intended to guide the Royal Society's role and shape its public statements on issues ranging from the efficacy of prayer to the nature and role of women must be introduced and voted on during council meetings.

YOUR ROLE IN THE ROYAL SOCIETY

As a Fellow of the Royal Society (FRS) and a member of the Executive Council, you are considered a leader of the organization. The Society has been in existence since 1660 and has counted among its members the most distinguished English scientists since. The nature of the organization is stated in the *Statutes of 1663*, which established the society under the patronage of the king:

> The business of the Society in their Ordinary Meetings shall be to order, take account, consider, and discourse of philosophical experiments and observations; to read, hear and discourse upon letters, reports and other papers regarding philosophical matters; as also to view, and discourse upon, rarities of nature and art; and thereupon to consider, what may be deduced from them, or any of them; and how far they, or any of them, may be improved for use or discovery (Charter Book of the Royal Society, RS manuscript DC/3, 1663).

The Royal Society sponsors lectures on a number of scientific topics. In 1829, the Earl of Bridgewater provided in his will for the president of the Society to enlist eight individuals to prepare a series of papers on the intersection between theology and natural science, known as the *Bridgewater Treatises*. The purpose of these treatises, inspired by William Paley's *Natural Theology; or, Evidences of the Existence and Attributes of the Deity* (1802), was specifically to instruct others:

On the Power, Wisdom and Goodness of God as manifested in the Creation illustrating such work by all reasonable arguments as, for instance, the variety and formation of God's creatures, in the animal, vegetable and mineral kingdoms; the effect of digestion and thereby of conversion; the construction of the hand of man and an infinite variety of other arguments; as also by discoveries ancient and modern in arts, sciences, and the whole extent of modern literature (Printed as a frontispiece to each of the treatises, from the Earl's last will and testament, 25 February 1825).

The *Treatises* appeared over the next eleven years and became the bedrock supporting the theological and scientific paradigm of natural theology. The eight treatises include: (1) "The Adaptation of External Nature to the Moral and Intellectual Constitution of Man" by Thomas Chalmers (1833); (2) "Chemistry, Meteorology, and Digestion" by William Prout, M. D (1834); (3) "History, Habits, and Instincts of Animals" by William Kirby (1835); (4) "The Hand, as Evincing Design" by Sir Charles Bell (1837); (5) "Geology and Mineralogy" by Dean Buckland (1837); (6) "The Adaptation of External Nature to the Physical Condition of Man" by J. Kidd, M. D (1837); (7) "Astronomy and General Physics" by Dr. William Whewell (rhymes with dual) (1839); and (8) "Animal and Vegetable Physiology" by P. M. Roget, M. D. (1840). As can be seen, the connection between science and religion encompassed every discipline. Science provided proof of the existence of a deity and science provided the clues necessary to understand the nature of that deity. The learned individuals who belonged to the Royal Society in the first half of the nineteenth century thus found no contradiction, but, instead, a mutual support among theology and the various disciplines.

Although there was no noticeable conflict between science and religion within the Royal Society in its early days, the class structure of English society threatened to prevent promising individuals from having adequate funding to pursue their research. Wealthy men who were interested in scientific discovery were frequently invited to membership because of the potential benefits of their influence and financial support, even if those men had exhibited no scientific ability themselves. In 1847, however, qualification for membership in the Royal Society began to change. No longer open to individuals whose wealth alone made them potential patrons of science, Royal Society membership was now based increasingly on merit. Thus, by the mid-nineteenth century the nature of the organization began changing from a social group with common interests to a professional academic organization.

As a member of the Royal Society, you will join your fellow members frequently for the Ordinary Meetings in which you present research findings. The organization meets formally on an annual basis for the Anniversary Meeting at which time awards and medals are conferred on members for their distinguished work. The

Charles Darwin, the Copley Medal, and the Rise of Naturalism

decision to award any individual is made by the Council. As a member of the Council, it is your responsibility to familiarize yourself with the work of the nominees, to evaluate the scientific merits of this work and to vote according to your principles.

THE COPLEY MEDAL

The Copley Medal is the most prestigious of all the awards given by the Royal Society. It is awarded annually for outstanding achievements in scientific research. Given only to living scientists, but not restricted in terms of nationality or subject, in recent years the Copley has alternated between the physical and natural sciences and has been awarded to Richard Owen (1851) for his work in paleontology, to Germans Alexander Humboldt for physics (1852) and Johannes Müller (1854) for physiology, and to Charles Lyell (1858) for geology, among others. The great American paleontologist Louis Agassiz won the Copley Medal in 1861 based primarily on his work on fossil fishes, but also in recognition of his broader geological theories.

In addition to granting these prestigious awards, the Royal Society Council is responsible for providing funds for scientific studies and for issuing public statements of support for various endeavors and policies that influence, encourage, or derive from scientific study. The Royal Society also sponsors lectures on specific topics that may include significant financial remuneration for the lecturer and that may reflect the Royal Society's position on various public policies.

During the early 1860s, Darwin's theory of natural selection began to influence the thinking of many members of the Royal Society, not only regarding the explanation for speciation, but also in many other areas. Explanations for "the way things are" began to shift from natural theology toward natural science. As a member of the Royal Society Council, you have an interest and an obligation to understand these explanations and to argue for what you believe to be true. You must evaluate the quality of the science and you must face the broader implications of your conclusions. Recall the Royal charge to the Society quoted above: the many discoveries of this amazing century will challenge everyone's way of understanding the world.

SPECIAL ROLES IN THE COUNCIL OF THE ROYAL SOCIETY

Gamemaster: Although the Gamemaster technically does not have a seat on the Council, the professor who plays the Gamemaster has the authority to make rulings on the play of the game or even make changes to rules and roles, as necessary. The Gamemaster may act as parliamentarian; any questions or controversy related to procedure should be appealed to him or her.

Preceptor: You may or may not have a student preceptor in your game. The preceptor also does not have a seat on the Council, but the Gamemaster may choose to use an experienced "Reacting" student as an advisor to the players and resource both in and out of game sessions. The preceptor may give advice during the play of the game, may meet with Fellows upon request, may help with resources, etc., and must not betray hidden agenda or factional goals. The Gamemaster should announce at the beginning of the course the exact function of this role.

President: General Sir Edward Sabine presides over all meetings of the Royal Society Council. General Sabine is responsible for ensuring that members follow parliamentary procedure during meetings, for ensuring that resolutions are appropriately presented and debated, and that members follow the agenda that has been established. General Sabine is free to use parliamentary procedure as he sees fit; however, he is expected to exhibit basic standards of fairness. He is free, and will be expected, to take an active and partisan position on some of the debates, always keeping in mind his role as president.

Secretary: A secretary has been designated (on his or her role sheet) for each of the two factions. The secretaries are responsible for establishing the agenda based on the items presented by fellow Council members. The secretaries, in collaboration with the Gamemaster, shall establish a means for compiling and distributing meeting agendas, making sure they are readily available for all players in a timely fashion.

Sergeant at Arms: At the end of each meeting, the sergeant at arms is selected by lot for the following meeting. The sergeant at arms is responsible for opening the meetings of the Council, leading the singing of "God Save the Queen," making any pertinent announcements and news, and keeping order during the meetings. The sergeant at arms serves the President and may be asked to fulfill appropriate tasks for the President, such as contacting the Gamemaster or finding the answers to questions through research. The sergeant at arms is also responsible for counting Council votes and verifying it with the Gamemaster.

SPECIAL RULES

RETENTION OF SEAT ON THE COUNCIL

At the outset of the game, all members have a vote in the Royal Society Council. This rule recognizes each Fellow's rhetorical and leadership abilities, the abilities that garnered a seat on the Council in the first place. Fellows who are members of either faction are expected to speak **at least** once each week (once per two meetings) in the Council. Any faction member who **fails** to speak at least once a week may lose voting privileges.

Indeterminates are under no such obligation, except to speak on topics of special interest to their characters. Some indeterminates **are required** to give papers and speeches at general meetings of the RS. (Check your individual role sheet.) Indeterminates are encouraged, at the very least, to ask questions and seek clarification of confusing points.

The Gamemaster will advise Fellows who are in danger of losing voting privileges. To register a voice vote or pose a simple question will not qualify as "speaking." Making a substantive comment while seated **does** qualify as speaking, as does, of course, making a speech or delivering an extended question at the podium. All Fellows have the right to approach the podium, so as to ensure their opportunity to speak. The Gamemaster will prod the Council President to ensure that everyone at the podium is recognized, sooner or later.

DISQUALIFICATION FOR READING ALOUD

Reading aloud is rarely an effective rhetorical strategy. Often it is boring and unsuited to an impressive leader such as you. Fellows are permitted to read aloud their **first** speech, but should not read subsequent speeches. Fellows are encouraged to consult 4x6 cards or notes.

ROLE OF GAMEMASTER, CONTACT WITH INSTRUCTOR

The Gamemaster's role is to do everything possible to make the game an intellectually broadening exploration of mid-nineteenth century English intellectual society. The game, accordingly, is complex. As the myriad elements of the game collide in innumerable permutations, chance will intervene in ways that the Gamemaster cannot anticipate. If the game careens wildly from historical plausibility, the Gamemaster may intervene, perhaps by modifying the rules or roles.

Most roles are difficult and challenging, as are the accompanying texts. Fellows should not hesitate to discuss their problems and confusion with the Gamemaster. The Gamemaster will clarify the best way to contact her or him. If, by the third week, you have not spoken privately with the Gamemaster (or the instructor) or exchanged at least one e-mail, you are probably playing the game poorly. E-mail is often a good way to initiate queries because the process of formulating a question in words helps clarify your thinking. The Gamemaster, in responding to such requests, will formulate a helpful response, keeping in mind the balance of what may and may not be discussed when advising Fellows with different roles.

Remember, the Gamemaster knows far more than you do about the game, including its permutations, complexities, and evolving modifications. In responding to particular queries, the Gamemaster will try to see the situation solely from the perspective of the questioner, without reference to what other players are doing or may likely do in the future. This is a way of saying that the Gamemaster will do his or her best to refrain from sharing your views and strategies with anyone else; conversely, you cannot assume that the Gamemaster will alert **you** to someone else's strategy or proposed initiatives.

STUDENT-INITIATED RULE MODIFICATIONS

Fellows may find that the constraints of the game (definition of roles, allocation of powers, description of victory objectives) are not consistent with their understanding of mid-nineteenth century England. Such Fellows are encouraged to appeal to the Gamemaster for a change in the rules or the addition of new ones. To that end, you may wish to conduct research on the subject and present it to the Gamemaster, perhaps as part of your written work for the game. The Gamemaster may or may not be persuaded by your brief; he or she may or may not change the rules. You may or may not be informed in advance of the Gamemaster's decision. But it is fair to assume that the Gamemaster will appreciate thoughtful research.

A WORD ON THE USE OF PROPS

Fellows often find it helpful in establishing their roles if they use some simple props that they associate with being in character. Such props may include simple costume additions, such as a top hat or an academic gown, or they may be something subtler like an (unlit!) cigar or spectacles. General Sabine may find it particularly helpful to have a gavel for calling the meetings to order. Whichever props you use should be appropriate to the time, setting, and character.

THE MAIN FACTIONS

Roles fall into three categories: A-Men, X-Men, and Indeterminates. Indeterminates do not have an immediate or obvious position on Darwin's receiving the Copley Medal, but may have deep convictions concerning one or more related issues of the time. They may be persuaded to join either side of the Copley Medal debate, but they have victory goals that are independent of that vote.

A-MEN (OPPONENTS OF DARWIN)

The A-Men essentially constitute the anti-Darwin/ pro-Natural Theology faction. They include General Edward Sabine (President of the Royal Society), Sir Richard Owen (the scientific leader of the faction), and an Anglican Bishop. There are also one or two members of this faction who are not based on explicit historical figures, but who help in the goals of the faction. In their arguments against Darwin, the A-Men may attack his failure to recognize the necessary consistency between Holy Scripture and science, his failure to provide adequate evidence for his arguments, or his failure to adhere to strict empiricism by conjecturing his own hypothesis regarding natural selection. Bishop Wilberforce's review of Darwin's theory criticizes it for many flaws, including inaccurate observations:

> We think it difficult to find a theory fuller of assumptions; and of assumptions not grounded upon alleged facts in nature, but which are absolutely opposed to all the facts we have been able to observe (1860. See full text of all quotes from Wilberforce's review in Appendix B).

It is **extremely unlikely** that the A-Men would argue for a strictly literal belief in the Bible; rather, they would argue that the fundamental truths of the Bible should help one understand the meaning of scientific discoveries. In fact, Wilberforce specifically refrains from religious arguments in his review:

> Our readers will not have failed to notice that we have objected to the views with which we have been dealing solely on scientific grounds. We have done so from our fixed conviction that it is thus that the truth or falsehood of such arguments should be tried. We have no sympathy with those who object to any facts or alleged facts in nature, or to any inference logically deduced from them, because they believe them to contradict what it appears to them is taught by Revelation. We think that all such objections savour of a timidity which is really inconsistent with a firm and well-instructed faith. . . . He who is as sure as he is of his own existence that the God of Truth is at once the God of Nature and the God of Revelation, cannot believe it to be possible that His voice in either, rightly understood, can differ, or deceive His

creatures. To oppose facts in the natural world because they seem to oppose Revelation, or to humour them so as to compel them to speak its voice, is, he knows, but another form of the ever-ready feebleminded dishonesty of lying for God, and trying by fraud or falsehood to do the work of the God of truth. It is with another and a nobler spirit that the true believer walks amongst the works of nature.

The A-Men accept the sufficiency of natural theology as the basis for explaining all natural phenomena. In light of this belief, they support the ideas stated in the Students' Declaration, a petition circulated by a group of Oxford chemists and chemistry students and eventually presented to the Lower House of the Convocation of Canterbury. Note that the Students' Declaration, like Bishop Wilberforce, insists that any incompatibility between science and religion is the result of a misunderstanding on the part of the scientist.

Declaration of Students of the Natural and Physical Sciences

We the undersigned Students of the Natural Sciences desire to express our sincere regret that researches into scientific truth are perverted by some in our own times into occasions for casting doubt upon the Truth and Authenticity of Holy Scriptures. We conceive that it is impossible for the Word of God, as written in the book of nature, and God's word written in Holy Scripture to contradict one another, however much they appear to differ. We are not forgetful that Physical Science is not complete, but is only in a condition of progress, and that at present our finite reason enables us only to see as through a glass darkly; and we confidently believe, that a time will come when the two records will be seen to agree in every particular. We cannot but deplore that Natural Science should be looked upon with suspicion by many who do not make a study of it, merely because of the unadvised manner in which some are placing it in opposition to Holy Writ. We believe it is the duty of every scientific student to investigate nature simply for the purpose of elucidating truth, and that if he finds that some of his results appear to be in contradiction to the written word, or rather to his own interpretation of it, which may be erroneous, he should not presumptuously affirm that his own conclusions must be right, and the statements of Scripture wrong; rather, leave the two side by side till it shall please God to allow us to see the manner in which they may be reconciled; and, instead of insisting upon the seeming difference between Science and the Scriptures, it would be as well to rest in faith upon the points in which they agree (1864).

An anonymous publication from the 1840s (later identified as the work of intellectual Robert Chambers) had previously introduced an idea similar to Darwin's that perhaps species changed over time, but that particular work denied any incompatibility between science and religion:

> I cannot but here remind the reader of what Dr. Wiseman has shewn so strikingly in his lectures, how different new philosophic doctrines are apt to appear after we have become somewhat familiar with them. Geology at first seems inconsistent with the authority of the Mosaic record. A storm of unreasoning indignation rises against its teachers. In time, its truths, being found quite irresistible, are admitted, and mankind continue to regard the Scriptures with the same respect as before. So also with several other sciences. Now the only objection that can be made on such ground to this book, is, that it brings forward some new hypotheses, at first sight, like geology, not in perfect harmony with that record, and arranges all the rest into a system which partakes of the same character. But may not the sacred text, on a liberal interpretation, or with the benefit of new light reflected from nature, or derived from learning, be shewn to be as much in harmony with the novelties of this volume as it has been with geology and natural philosophy? (*Vestiges of the Natural History of Creation, 1844*)

The A-Men applaud this work for its application of natural law to social as well as natural sciences, but they also strictly adhere to its belief that an accurate understanding of natural law must be in accordance with the Scriptures. On the other hand, they are cautious about interpreting scientific theories in accordance with divine revelation because scientific results may be quickly overturned. As Wilberforce wrote in his review:

> Few things have more deeply injured the cause of religion than the busy fussy energy with which men, narrow and feeble alike in faith and in science, have bustled forth to reconcile all new discoveries in physics with the word of inspiration. For it continually happens that some larger collection of facts, or some wider view of the phenomena of nature, alter the whole philosophic scheme; whilst Revelation has been committed to declare an absolute agreement with what turns out after all to have been a misconception or an error. We cannot, therefore, consent to test the truth of natural science by the Word of Revelation. But this does not make it the less important to point out on scientific grounds scientific errors, when those errors tend to limit God's glory in creation, or to gainsay the revealed relations of that creation to Himself. To both these classes of error, though, we doubt not, quite unintentionally on his part, we think that Mr. Darwin's speculations directly tend.

The A-Men will thus attack Darwin's work as it "limits God's glory in creation," but they will also attack it on scientific grounds and because of its departure from pure empiricism to speculation.

X-MEN (SUPPORTERS OF DARWIN)

The X-Men are members of a group that will emerge more formally in 1864 as the X Club, which formed in explicit opposition to the religiously conservative signers of the "Declaration of Students of the Natural and Physical Sciences." The X-Men hold a more naturalistic position and include character roles for Thomas Henry Huxley (Darwin's Bulldog), Joseph Hooker (Darwin's Confidante), and a chaplain/tutor associated with the Royal Family. There are also one or two additional members of the X-Men faction who will work for the goals of the group.

The X-Men argue against both Chambers and the Students' Declaration, insisting that science must not be held back by doctrinal limitations. They insist that [if one begins to look at the empirical evidence without the established prejudice of religious training, one will find scientific laws that operate without divine interference.] One of the X Club members, John Tyndall, would later clearly describe the application of this practice to Darwin's theory of natural selection.

> The change, however, from form to form was not continuous, but by steps—some small, some great. 'A section,' says Mr. Huxley, 'a hundred feet thick will exhibit at different heights a dozen species of Ammonite, none of which passes beyond its particular zone of limestone, or clay, into the zone below it, or into that above it.' In the presence of such facts it was not possible to avoid the question: —Have these forms, showing, though in broken stages and with many irregularities, this unmistakable general advance, been subjected to no continuous law of growth or variation? Had our education been purely scientific, or had it been sufficiently detached from influences which, however ennobling in another domain, have always proved hindrances and delusions when introduced as factors into the domain of physics, the scientific mind never could have swerved from the search for a law of growth, or allowed itself to accept the anthropomorphism which regarded each successive stratum as a kind of mechanic's bench for the manufacture of new species out of all relation to the old.

> Biased, however, by their previous education, the great majority of naturalists invoked a special creative act to account for the appearance of each new group of organisms. Doubtless there were numbers who were clear-headed enough to see that this was no explanation at all, that in point of fact it was an attempt, by the introduction of a greater difficulty, to account for a less. But having nothing to offer in the way of explanation, they for the most part held their peace. Still the thoughts of reflecting men naturally and necessarily simmered round the question ("The Belfast Address," first presented in 1874, appeared in *Fragments of Science: A Series of Detached Essays, Addresses, and Reviews, vol. 2, New York: D. Appleton, 1896*, pp. 171-72).

In sympathy with this position is T.H. Huxley, who not only wishes to free science from the shackles of religious doctrine, but also wishes to increase the professionalization of scientific research. Huxley views science as a field for experts, rather than amateurs, and believes that theologians should stick to their own area of expertise, rather than trespass into the specialized fields of natural and physical science. He advocates a religious position for which he later coined the term "agnosticism," by which he implies that one should approach religious questions with the same degree of skepticism with which one approaches other fields of knowledge. No assumptions should be accepted as absolute and all areas of understanding should be open to investigation and question. In addition, Huxley sees science as a means to establishing a broader and more equitable political and economic system. Perhaps most importantly, Huxley views Darwin's notion of natural selection as the best available explanation for the existence of species and champions it as "simple, obvious, and brilliant."

The members of the X-Club are a network of men who will have a pervasive and lasting influence on the practice and professionalization of science throughout the Commonwealth and in the United States. The nine formal members of the club include three physicians-turned-naturalists (Huxley, Hooker, and Busk), three engineers-turned-physicists (Tyndall, Frankland, and Hirst), two well-connected young gentlemen naturalists (Lubbock and Spottiswoode), and the outspoken speculative philosopher Herbert Spencer, who connects the group to London's literary circles. Huxley and the three physicists have worked their way up to positions of influence solely on the basis of their own intellectual merit and, consequently, are strong advocates of the professionalization of science. By the time our game begins, Hooker is Assistant Director of the Royal Gardens at Kew and Busk has succeeded the renowned Richard Owen as the Hunterian Professor of Comparative Anatomy at the Royal College of Surgeons, thus providing the group with substantial scientific credibility. Meanwhile, the strategic addition of the young Anglican gentlemen John Lubbock, son of an influential London banker, and William Spottiswoode, son of the Queen's printer, further solidifies the broader cultural power of the group.

The X-Men will argue strongly in favor of Darwin because of his impeccable research, his willingness to speculate without deference to theological considerations, and because of their personal support for him as a fellow scientist.

BRIEF SKETCHES OF GAME CHARACTERS

HISTORICAL AND COMPOSITE FACTIONAL CHARACTERS

A-Men Faction

General, Sir Edward Sabine: One of the "grand old men" of the scientific community. Born in Dublin, Ireland, in 1788, and educated at the Royal Military Academy in London, he is a British military officer first and foremost. He traveled to the Arctic with William Parry and around the world more than once. Trained as an astronomer, he devised copious experiments using magnetism and pendulum measurements to determine the size and shape of the Earth. He served as a secretary of the Royal Society beginning in 1827, was vice-president of the organization from 1850-1861, and has just been elected president.

Sir Richard Owen: Born in Lancaster, England, Owen grew up in a relatively poor family. He served as a midshipman in the Royal Navy, but preferred science and began training at the University of Edinburgh. Eventually, he became a surgeon licensed through the Royal College of Surgeons in London, but his real interest is in comparative anatomy, rather than medicine. Owen received the Copley Medal in 1851, served on a series of government committees, and acts as an advisor and expert witness to the government on all manner of scientific matters.

Bishop in the Anglican Church: Fifty years old at the time of the 1860 Oxford Debate, the bishop has a national reputation for handling controversy well. For years he has allied himself with Samuel Wilberforce, the Bishop of Oxford, holding the middle ground against the Oxford Movement, a conservative doctrinal effort to remove the distinction between the "High" Anglican Church and the Catholic Church. They both have been diligent in their efforts to hold on to those who have been drifting toward Catholicism, while maintaining the highest standards for clergy throughout England. The bishop is also a trained botanist and loves breeding roses.

X-Men Faction

Thomas Henry Huxley, "Darwin's Bulldog": Born in 1825 to a struggling math teacher and a Cockney mother, Huxley is sixteen years younger than Darwin. As a working class Irishman, he is considered a bit coarse for upper crust British society. He is, however, eloquent and argumentative, and he has little patience for slow-witted fools and intellectually lazy or dishonest people. Following Huxley's confrontation with Bishop Wilberforce at the 1860 meeting of the BAAS, he became known as "Darwin's bulldog," a nickname he has come to accept and even to relish. In his book, *Zoological Evidence on Man's Place in Nature* (1863), he set out a comprehensive review of what is known about paleontology and ethnology of humans and other primates.

Joseph Dalton Hooker, "Darwin's Confidante" (Optional for larger game): Botanist, biographer, and world traveler, Hooker is the youngest son of Sir William Jackson Hooker, former professor of Botany at Glasgow University and current Director of the Royal Botanical Gardens at Kew (since 1841). Born in 1817, he began attending his father's university lectures on a regular basis at the age of seven and grew up immersed in botany. He graduated from Glasgow in 1839 with a degree in medicine.

Chaplain/Tutor: A theological supporter of Darwin's views and an associate of Charles Kingsley, he believes that the Deity "created primal forms capable of self development into all forms needful" and that such a view inspires more reverence than the belief that each species represents a fresh act of creation. Supportive of reform efforts in the Anglican Church and sympathetic to the views represented in *Essays and Reviews,* his scientific expertise is in botany.

Additional Faction Member: Depending on the number of players in your game, factions may include additional members who will develop their own characters in consultation with the Gamemaster.

INDETERMINATE ROLES

Anthropologist: Thirty-seven years old in 1862, he hails from a modestly well-to-do family. The anthropologist grew up in Edinburgh, the son of a factor, a merchant/agent who owned five ships engaged in Atlantic trade. After earning a degree in medicine, he traveled the world, including a trip with Richard Burton to Africa, boating up the Nile to its source at Lake Tanganyika. The anthropologist is especially familiar with the work of Louis Agassiz.

Astronomer: Sixty-five years old in 1862, the astronomer is a well-established gentleman. A founding member of the British Association for the Advancement of Science and an FRS for fifteen years, he inherited a substantial income from his father. The astronomer married into a highly regarded family of minor nobility and settled into a country estate outside of London. His wife is witty and brilliant, making them a popular couple during the social season. The astronomer loves to talk about his field, in particular, and scientific matters, in general.

Chemist: A bit stodgy, the chemist has long been active in scientific societies and is a great friend and colleague of Herbert McLeod. The son of a physician, he considered a career in the military but decided to attend the Royal College of

Chemistry. Having recently turned sixty, the proud grandfather of six, he has become increasingly aware that he is a little on the rigid side. He might even be proud of it. Still, the chemist is known to have a good sense of humor and pokes fun at himself when the situation warrants.

Civil Engineer: At the age of forty-five, the engineer has several professional acquaintances in the RS. His father was a master carpenter in Manchester, a center of industry, science, and host to several scientific organizations. He attended King's College, one of the first institutions of higher learning to teach engineering. He began his career working for the Manchester and Bolton Railway and eventually joined not only the RS, but the British Association for the Advancement of Science, and later the more elite Manchester Literary and Philosophical Society. Like the chemist, the engineer has a sense of humor, but his tends more toward the satirical.

Geologist: As an active member of the British Association for the Advancement of Science, he is well-read in science journals and was, of course, present for the Huxley-Wilberforce debate. Thirty-two years old, and thus relatively young, he is considered a rising star among geologists. His great-grandfather was a loyalist plantation owner in South Carolina, who fled North, then migrated to England after the King's troops evacuated New York City at the end of the American Revolution.

Historian: A highly revered and quite elderly classicist, the Historian is renowned for his keen mind and incisive questions. His positions on scientific matters are a bit impenetrable, but no one doubts his intellectual acumen. His relentless and probing questioning can make even the most self-assured Council members uncomfortable at times. No one is exempt from his Socratic queries.

Mathematician: Born in Glasgow in 1813, the mathematician is the only son of a prominent merchant. Enrolled at Glasgow University at 15, he distinguished himself in classics, mathematics, and physics. From there he moved on to Trinity College, Cambridge where he was elected a fellow in 1836. He is proud of his family history and has ties to the Royals. He wrote a number of influential scientific papers related to the algebra and analytical geometry of optics and co-founded the *Cambridge Mathematical Journal*. The mathematician just completed a large monograph entitled *Admiralty Manual for Ascertaining and Applying the Deviations of the Compass Caused by the Iron in a Ship*.

Mineralogist: The mineralogist has worked for the Royal Geological Survey for thirty years and has many friends among mining engineers, leading industrialists, and important government officials. Primarily interested in work related to coal deposits, he keeps up on the literature in a broad array of scientific fields. By joining scientific organizations, he maintains contacts and friendships from Trinity College. Edward Sabine is one of his oldest friends.

Charles Darwin, the Copley Medal, and the Rise of Naturalism

Naturalist and Ethnologist: One of the youngest members of the Royal Society, he was elected based as much on family ties as on demonstrated scientific merit, though he is clearly also a rising intellectual star. His father is a central figure in the British banking establishment with close connections to the Royal Family. He is fervently committed to many liberal causes, such as the universal abolition of slavery; liberalization and reform of the Anglican Church, expanded rights for women, as well as ethnic and religious minorities; increased educational and recreational opportunities for the working class; and social programs aimed at reducing the suffering of the poor. He is also a member of the Ethnological Society, which is currently engaged in a very public controversy with the upstart Anthropological Society of London, a group dedicated to promoting a "scientific" theory of race.

Paleontologist: A follower of Charles Lyell, the paleontologist has vast experience in field work, having traveled as much as possible in his quest for unusual fossils. Also an expert at comparative anatomy, he holds the position of Lecturer in the Natural Sciences at Downing College, Cambridge. He is a member of the Geological Society, the BAAS, and the Royal Society. As an ardent fossilist, he is particularly interested in Darwin's fossil finds.

Philosopher, Friend of Mill: This philosopher is a long-time student and friend of John Stuart Mill—an eminent British social philosopher and logician—and his thinking has been heavily influenced by Mill. He frequently boasts of this connection, but most members of the RS take this name dropping with good humor, especially since the philosopher has a very sharp mind and is quick to point out logical fallacies in any arguments put before him. He is more interested in philosophical issues than in field work or experimentation.

Philosopher, Inductivist, "Disciple of Bacon": This philosopher loves to remind everyone in the Royal Society that Francis Bacon's thinking provides the basis for all good science and that Isaac Newton was the greatest scientist who ever lived. Occasionally his droning on and on about Bacon and Newton makes other members of the RS wonder if he thinks anything has happened at all in the scientific field during the last century. On the other hand, the inductivist is always a good fellow to have on your side in an argument. Whenever anyone speculates beyond the empirical evidence in a particular field, he is quick to point out that the scientific method ought to be limited to induction.

Depending on the number of players in your game, some of the roles noted above may be omitted.

Both General and Council Meetings will be conducted according to parliamentary procedure as explained in the handout. The President and the Sergeant at Arms are responsible for enforcing these rules. All Fellows should review and be familiar with parliamentary procedure as explained in *Robert's Rules of Order*. Fellows are encouraged to purchase abbreviated forms of *Robert's Rules of Order* to use during the game. Information about these rules may be found at http://www.robertsrules.org.

Each meeting will have an agenda that includes those items required by the game and those added by Fellows. The president may impose time limits on issues, but some free discussion should be encouraged. There is one limitation to the powers of the President in the form of the podium rule.

PODIUM RULE

Fellows, if they wish, may speak while seated, but it is important that every Fellow go to the podium to make a more formal statement or speech at least once during the game. Also, Fellows who find that they are not being recognized by the President owe it to themselves, their faction, the instructor, and the class as a whole to approach the podium. The President is expected, at some reasonable point, to call upon the person at the podium. If one person is already waiting at the podium, others may join a line behind him to speak. If necessary, the Gamemaster will remind the President of the need to allow those at the podium to speak.

COPLEY NOMINATIONS

Nominations for the Copley Medal will be made during the penultimate session (see below). Be aware that Council members are excluded and may not be nominated either by themselves or others to receive the award.

PRELUDE TO THE GAME

During the class session prior to beginning the game, Fellows will be assigned character roles and may view a video re-enactment of the 1860 Oxford Debate. Following the video, the Agenda for the first Council Meeting should be reviewed and brief faction meetings should be held. Factions should also plan to meet outside of class to prepare strategies for the first round of game play. The Sergeant at Arms for the first Council Meeting should be chosen by lot.

Indeterminates may also wish to meet together at this time and should meet individually with the Gamemaster and/or preceptor. It is critically important that characters with indeterminate roles not reveal too much about their specific victory objectives early in the game. Any person who is giving presentations at general meetings should consult with the Gamemaster as soon as possible. You may refer to the game web site at http://www.darwingame.org for various Internet links that may help you with your research.

DETAILED AGENDA (SESSION BY SESSION)

The following sessions are set for all versions of the game.

Session 1

General Session

General Sabine will open the first session by calling for the Sergeant at Arms to lead singing of "God Save the Queen" in its entirety. At subsequent General Meetings, at least two verses must be sung. The selected two will be at the discretion of the sergeant at arms.

Sabine will read Prince Albert's Obituary and comment. He will then recognize the chaplain/tutor first, then the Bishop (Natural Theologian), and then anyone else who may wish to speak on the importance of Prince Albert's life and death.

At the conclusion of these remarks, the general meeting will be adjourned.

Council Session

Sabine calls the Council meeting to order and calls for any additional discussion about the death of Prince Albert and, specifically, the tension between science and religion. A public argument over the effectiveness of prayer versus reliance on scientific medicine became quite acrimonious during the Prince Consort's illness. After Albert's death, blame began flying in both directions.

To prepare, read the "Introduction to Victorian England" and "Religion in Great Britain" sections in the game book very carefully.

Session 2

<u>General Session</u>

Overview of Darwin's key ideas.

Philosophy of Science presentations.

<u>Council Session</u>

Probable resolution regarding criteria for judging "good scientific work."

Sessions 3-5

May consist of General and/or Council meetings as needed and agreed upon.

Session 6

<u>Council Session only</u>

Formal nominations for the Copley Medal and additional business.

Session 7

<u>Council Session only</u>

Debate on Copley Medal, voting and any remaining business.

Session 8

Award Dinner.

SUMMARY TABLE OF AGENDA AND ASSIGNMENTS:

Session	General Meeting	Council Meeting
Pre-Game I: Historical Context & Role Assignments	None	None
Pre-Game II: Huxley-Wilberforce Film and Initial Faction Meetings	None	None
Game Session 1	Prince Albert's Obituary Chaplain's Comments/Bishop's Prayer	Huxley's Efficacy of Prayer Resolution (Chemist's Counter-Resolution?)
Game Session 2	Hooker's Overview of Natural Selection (Plus philosophers' comments)	Nature of science Resolution supporting role of speculation
Game Session 3	Essays & Reviews | Student's Declaration (Chaplain Tutor/Chemist)	Resolution of Support for authors of essays and reviews (Chaplain/Tutor)
Game Session 4	Scientific Views on Race and Society Joint presentation by Anthropologist & Ethnologist	Condemnation of Anthropological Society (Ethnologist) Study of Racial Variation Proposal (Anthropologist)
Game Session 5	Women and/or Public Health Malthusian versus Engineer	Pro-woman resolution or nomination Public Health
Game Session 6 [Copley Nominations]	None	Copley Nominations (additional new or old business as time allows)
Game Session 7 [Copley Vote]	None	Copley Debate and Voting (old business only if time allows)
Game Session 8 [Anniversary Meeting & Post-game Analysis]	Anniversary Meeting	None

Additional Agenda Items to be Debated Prior to Final Council Meeting

Resolutions should be submitted for debate and Fellows will vote on each of the items below, although Fellows may introduce other resolutions and proposals not on this list. Debates should include appropriate references to Natural Theology and naturalism as they apply to these issues. Specifically, members should focus on the implications of Darwin's evidence throughout these debates. Agenda items include:

The use of scientific information to improve public health;
Nominating women (or a specific woman) to the ranks of the Royal Society;
Discussion of the scientific evidence on the nature of race;
Supporting or denouncing the "Declaration of Fellows of the Natural and Physical Sciences" (see text above); and
Supporting or denouncing the authors of *Essays and Reviews*.

Agenda for the Final Two Sessions of the Game

The agenda for the penultimate session must include the official nomination, debate, and vote for the winner of the Copley Medal. Specific characters are assigned to make certain nominations; however, other characters may make their own nominations. Nominations must be made formally, with specific justifications for why that individual deserves the award. All nominations must be seconded by another member of the Council in order to be considered in the final vote. You should be familiar with all the nominees in order to participate in the debate. Debate may include praise or criticism, but must follow the behavior expected of gentlemen. Remember that this is your opportunity to persuade the indeterminates.

Although voting is to be by secret ballot, ballots must be signed because your second paper will be a justification of this vote. After the voting occurs, the Council will recess while the President and the Gamemaster tally the vote. The winner must have a simple majority (50% + 1). If there is no majority, the President will reconvene the Council and announce the results. The President and the Council will then discuss the best way to achieve a majority vote. The President may consult with the Gamemaster to determine the best course of action. If there is a clear majority, then the President adjourns the meeting, but does not announce the winner.

The Royal Society Anniversary Meeting Dinner will be held at the last session. [Food is recommended!] After dinner, the President will formally announce the winner of the Copley Medal and the Council's rationale for conferring the medal. The Gamemaster, or an appointed spokesperson, may read the original after-dinner speech given by Charles Lyell. The Gamemaster will then conduct a debriefing that will include a historically accurate account of events leading to Charles Darwin's award of the Copley Medal for 1864.

PROTOCOL AND PARLIAMENTARY PROCEDURE

Sir Edward Sabine will preside over each meeting of the Royal Society and Council. A secretary will be appointed from each faction before the first Council meeting. Special instructions for secretaries are included on their role sheets. The secretaries, in consultation with General Sabine, will set the sequence and post the agenda for the session. The secretaries should establish a clear protocol for posting agenda items before the first session of the Council. Anyone wishing to place an item on the agenda **must** contact one of the secretaries.

Remember, secretaries may have factional interests! The Gamemaster reserves the right to intercede in the agenda in the interest of game progress. Agenda items may include resolutions, position papers, and other business pertaining to factional or individual victory conditions.

SPECIFIC WRITTEN AND ORAL ASSIGNMENTS

1. In order to qualify as a player of this game you may be required to demonstrate your understanding of the basic elements in Darwin's argument by passing a quiz.

2. Each student is required to present at least one formal oral argument to fellow Royal Society Council members. This presentation need not directly address Darwin and the Copley Medal, but should be developed within the context of the debate concerning Darwin's ideas. While this statement should not be read verbatim, a written summary of the argument must be given to the Gamemaster.

3. Each student will be required to write at least one position paper, focusing specifically on one of the pertinent issues, which will be made available to other Fellows. This paper need not be related to the oral presentation noted in Assignment 2. This work will be published either online on a course web site or in a tract distributed in hard-copy form to fellow Council members.

4. Fellows will be expected to maintain complete records of written correspondence with other players. Some of that correspondence may be public during the game. Private correspondence about the game should be preserved and will be collected as a part of the student's written grade. Fellows should visit the Darwin Correspondence Project web page to get a feel for the style of gentlemanly Victorian correspondence (http://www.lib.cam.ac.uk/Departments/Darwin/).

5. At the end of the game, each student must submit a four-to-five page synthesis explaining how Darwin's theory affected the positions your character took during the course of the debates in the Council and how your character might experience the paradigm shift that is occurring. You must include a justification of your character's vote on the Copley Medal.

6. Fellows **may** apply for the Royal Society Prize to Support Research. This application consists of a prospectus to conduct specific research in one of the fields of science. The paper must include a summary of the information known in the field as of 1864 and it must describe the methodology proposed by the researcher. The paper must also state whether the methodology follows the inductive scientific method or whether it is the testing of a specific hypothesis. If the researcher is hypothesis testing, the paper must specifically state the author's hypothesis and how it is to be tested. This paper is **not** a substitute for the required position paper, but is in addition to the required work. The proposals will be judged by a panel of Gamemasters. Only one paper will receive the Prize; however, all papers will receive credit depending on the quality of the work.

PLAYING A NATURAL PHILOSOPHER OR MAN OF SCIENCE IN VICTORIAN ENGLAND

Soon after the publication of *On the Origin of Species,* Darwin sent out one hundred copies of the book to friends and colleagues with the intent of earning the support of reputable individuals. He sent one copy to his old teacher, Adam Sedgwick. To Darwin's dismay, Sedgwick subsequently wrote in *The Spectator* that "Darwin's theory is not inductive—based on a series of acknowledged facts pointing to a general conclusion—not a proposition evolved out of the facts, logically, and of course including them. To use an old figure, I look on the theory as a vast pyramid resting on its apex, and that apex a mathematical point" (March 24, 1860). While bypassing Darwin's conclusions, Sedgwick aimed his criticism directly at the *method and structure* of the argument. What in Darwin's method did Sedgwick find objectionable? In *Charles Darwin, the Copley Medal, and the Rise of Naturalism,* you will be playing the role of a scientific authority. As a critical part of that role, your character is required to evaluate a scientific theory, much like Sedgwick did. The following section will help prepare you for that role.

What does it mean to evaluate a scientific theory? The question is a difficult one because the definitions of science and theory are never clear and they are always subject to change. One of the first people who engaged in scientific inquiry was the Greek philosopher Aristotle (382–322 B.C.). For Aristotle, science was a method for gathering knowledge by means of observation. This approach contrasted the methods employed by his teacher, Plato (428–327 B.C.), who believed that real knowledge came from pure reason and intuition. Even so, Aristotle's version of science was very different from science today. What science means changes as the boundaries of our knowledge expand and our limitations are better understood? Science is messy and there is still debate about what the scientific method *is*.

In science classes, you have probably encountered the scientific method: the recipe for investigation that has served as the essential tool used by the scientific community. This method ideally provides for observable verification, objectivity, and certainty. You were likely taught that the scientific method conforms to the following model. First, you observe; next you make a hypothesis; then experiment, collect data, and, finally, draw a conclusion through analysis. If your original hypothesis is confirmed by your own experiment and reconfirmed by other scientists' experiments, then your hypothesis eventually becomes a theory. Or maybe you were taught that it is impossible to create a hypothesis before one has gathered data. We assume that science is constructive, progressive, and perhaps inevitable, and that the scientific method prevents scientists from following the wrong path for very long. This view simplifies the situation and fails to account for the very real debates that have occurred regarding the appropriate methods of science.

In this game, your character needs to have some opinion about what science is and what counts as acceptable scientific method or you will be unable to effectively evaluate Charles Darwin's theory. Since science has changed since the nineteenth century, we should ask, "What is science to a Fellow of the Royal Society in the early 1860s?" How you answer this question will shape your opinion about whether or not Darwin merits the Copley Medal.

For centuries, natural philosophers have discussed the most reliable ways to conduct science. The types of questions these philosophers asked included: How is one supposed to make observations? How is one supposed to create a hypothesis? Once the hypothesis is made, how is one supposed to test the hypothesis? What is a "law" or principle of science? Are these laws reliable descriptions of the natural world's operations? Can we *know* anything with certainty in science?

One of the major points in dispute about the theory of natural selection concerned the *method* that Darwin employed. In other words, rather than argue that Darwin's theory was incompatible with the Bible or with some alternative scientific theory, many people (including members of the Royal Society) argued that Darwin made mistakes in observation or fell victim to logical fallacies when he offered explanations for his observations.

INTRODUCTION TO THE PHILOSOPHICAL CONTROVERSY

Logical Foundations of Inductive and Hypothetical-Speculative Scientific Methods

After about 2,000 years of Aristotelian science, a great scientific revolution, marked by Copernicus' heliocentric theory of the solar system, began in the sixteenth century. In *The Advancement of Learning* (1605) and *Novum Organum* (1620), Englishman Francis Bacon (1561–1626) laid down the philosophical foundations of the modern scientific method. He stressed that knowledge of the natural world does not come from the simple assertion of authority. Instead, knowledge comes from experience. Individual observations accumulate and, using this method, natural philosophers can *slowly* begin making well-justified generalizations and theories. Referred to as the inductive method, this new approach soon became the dominant and accepted method of the sciences throughout Europe.

Both the inductive method and another method, the hypothetical-speculative method, which will be discussed later in this section, emphasize a different form of logic at their core. Although both methods employ both inductive and deductive logic, they incorporate them differently.

Induction and deduction are the two primary forms of reasoning. Their origin in formal logic and science can also be traced back to Aristotle, who invented the first formal logic. The method of induction begins at particulars and moves to generalizations; its conclusions are probable, not necessary. Inductive reasoning is judged by its varying levels of strength and weakness. Deduction begins at generalizations and moves to particulars and its conclusions necessarily follow from its premises. If the premises are true, then a valid deductive argument necessarily results in true conclusions.

An inductive argument could start from observations of particular events and move to a general statement about all events of a similar kind. For example, suppose you are given a bag of marbles. You cannot see the marbles, but you are told that there are about one thousand marbles in the bag and that it is your task to determine what color the marbles are. You pull out one marble and you see it is black. At this point, you may draw the conclusion that all the marbles are black, but you must admit that you are probably jumping the gun. So, you draw out a second marble, which is also black. At this point, you have a stronger claim than before that the marbles are all black, but, while stronger, your conclusion is still objectively weak. You draw twenty more marbles from the bag and all of them are black. You shake the bag to make sure your sample is fair, draw twenty more, and they are still all black. At this point, you feel fairly confident that all the marbles are black, but, admittedly, you are not absolutely sure; there may be one white marble, or there maybe 958 white marbles. This reasoning is inductive: you moved from *particulars*, in this case a group of particular marbles, to a *general* statement, that *all* marbles in this bag are black.

In science, *induction is usually used to make predictions about unobserved cases*. With the bag of marbles, you have made predictions about the unobserved marbles. Inductive arguments are rarely fool-proof; they are always judged as strong or weak and this judgment is not an either/or, but rather one of degrees. By making a case based on one marble, your theory that *all marbles in this bag are black* was weaker than when you used two, while drawing forty-two makes your argument substantially stronger. The only time when an inductive argument is fool-proof is when *all* particulars have been observed. For example, if all one thousand marbles were observed and you drew the conclusion that all marbles are black, you have proven your inductive argument. This degree of certainty is usually impossible to achieve concerning the discovery of scientific laws, which is why it is usually impossible to call a scientific theory a fact.

Inductive logic is often used to make predictions about future events, which are a type of unobserved events. When asked if the sun will rise tomorrow, you may reply, that, yes, the sun will rise tomorrow. You know this because it has risen every day of your life. If this is your line of reasoning, then you have moved from a set of observed cases (in this case, the set of all days that you have lived) and made a prediction about an unobserved case, i.e., tomorrow.

Deductive logic works in the opposite direction: from generalizations to particulars. Now, suppose you are given a second bag of marbles, and you are told from a credible authority that *all the marbles in this bag are red*. Now, you pull a marble out of the bag and, without looking at it, you ask yourself, "what color is this marble?" Well, because *all marbles in this bag are red*, then *this particular marble that I hold in my hand must be red*. You look at the marble and, with no surprise, you see that the marble is red. This may not seem like a very powerful mode of reasoning, but it has its advantages. First of all, there was no doubt. If all marbles are red, then, without fail, any particular marble will be red. This certainty marks a clear advantage over inductive reasoning, where the vast majority of conclusions are merely possible and not proven.

You may have some suspicions that deductive arguments are not as fool-proof as they appear at first glance. Let's break down the marble deductive argument. The statement *all marbles in this bag are red* is like a law of nature. *Any particular marble in this bag is red* is like a conclusion about unobserved cases in the natural world, and opening your eyes to see the color of the marble is analogous to a test. Then, again, if you are using a sort of reasoning that is infallible, what is the purpose of testing? You already know that you are correct, right?

But suppose that after you have been told that *all marbles in this bag are red*, and after you have deduced that the particular marble that you hold in your hand is red, you then open your eyes to discover that the marble is, in fact, not red, but, rather, blue. What does this discovery mean? Surely, you reasoned correctly. *If all marbles in this bag are red, then any particular marble in this bag is red.* That reasoning is certain. By showing that there was a marble that was not red, however, you also show that the statement *all marbles in this bag are red* is false.

The problem with deductive reasoning in science is that it does not depend on observation of the natural world to come to conclusions. If a scientist were to begin from a generalization and deduce various conclusions from that generalization, it is quite likely that the conclusions will not be true. When a scientist makes predictions grounded in some *law of nature*, a law implies that anytime a certain set of conditions is fulfilled, a certain result *must* occur. The law is the generalization and particular cases are determined by the generalization, but if scientists discover that the predicted result does not occur, then they have revealed that the *law* is not a law at all.

All natural philosophers, including Aristotle, start from observations and then infer some generalization based on that observation. The difference between a good inductive theory and a poor one is that in a good one, the step of induction follows strongly from the observations, where the truth of the premise renders the truth of the conclusion highly probable. In a poor inductive theory, the initial observation may or may not be true and, thus, the reasoning from the initial premise to the conclusion is weak, which often results in an overgeneralization. Once the conclusion from induction is made and it is used as a law or theory, conclusions can be drawn predicting how things in the world will behave. So, the scientific process involves first a step of induction and then a step of deduction: we observe particulars to induce some general laws or principles of the natural world and then use those general laws or principles to deduce things about particular cases. This step from observation to generalization should occur when an appropriate number of observations have been gathered, a condition called a *consilience of induction*.

It often happens, however, that a scientist has made the generalization too quickly or that they have made observations and unwittingly made a generalization or interpretation of the data that did not follow from the observations (in other words, they introduced some information into the scientific process that lacked empirical basis). When a scientist constructs a theory based on false generalizations or assumptions, the scientist has almost inevitably made a gross error and can certainly be accused of following inappropriate methodology.

The major point of contention between the X-Men and A-Men is whether Darwin used the method of induction or deduction. If the X-Men can show that Darwin did use induction, then the X-Men can prove Darwin used appropriate methodology. If the A-Men can demonstrate that Darwin used only deduction from hypothesis, they can argue that Darwin was not engaged in real science, but rather metaphysical philosophy. There is general consensus that metaphysics, which used pure reasoning, rather than observation, was a wholly inadequate way to discover scientific truths.

Although Darwin's *On the Origin of Species* is packed with facts and observations, Sedgwick, an eminent scientist and one of the founders of the modern discipline of geology, claims Darwin's theory is not "based on a series of acknowledged facts pointing to a general conclusion." In other words, despite the fact that Darwin makes many observations, Sedgwick believes Darwin introduced a layer of unjustified assumptions into his observations. He argues that it is these assumptions that Darwin uses to construct his theory of natural selection, rather than the observations, themselves. Therefore, Darwin based his theory on a general idea, rather than a series of particulars and used deduction from hypothesis, rather than induction from observation. If Sedgwick is right, Darwin does indeed have some problems.

It will be your task to determine whether or not Sedgwick's charge regarding scientific methodology is justified (A-Men will support Sedgwick and X-Men oppose). Darwin, of course, denies this charge and, has written that his theory does in fact use induction. He argues in his typical understated fashion, "from [the] analogy of domestic productions, & from what we know of the struggle of existence & of the variability of organic beings, [it] is in some very slight degree in itself probable" (Darwin to J.S. Henslow, 8 May 1860, in *Correspondence of Charles Darwin*, vol. 8, p. 195). In short, Darwin points out that as his theory is supported by argument from analogy (one form of inductive argument) and by the two massive bodies of particular examples he accumulated that moved him toward generalization, he has certainly used the inductive method.

The Role of Verae Causae *or "True Causes" in Science*

One way of stating the basic mission of science is to say that it is the purpose of science to discover the causal link between specific events. For example, as one holds up a rock at shoulder height from a stationary position and lets go, one witnesses the rock fall directly towards the Earth. It is up to science to understand the cause of that event. Cause, however, can be difficult to determine.

You might say that the rock falls to the Earth "because" you let go of it. That is known as an immediate or proximate cause: the initial act that sets off a continuous sequence of events that leads to an event occurring, in this case, a rock falling. You might also say that the rock falls because you are too tired to hold it up any longer. That might be known as the efficient cause: an agent that causes an event to occur, in this case, you would be considered the agent. Or you might state that the rock falls because of the law of gravity, which is an absolute condition requiring "every object in the universe attracts every other object along the line of centers of the two objects that is proportional to the product of their masses and inversely proportional to the square of the separation between the two objects" (*Principia*, 1687). In other words, as stated in Newton's Law of Gravity quoted above, the rock falls because it is close to the Earth and the Earth is so much bigger that it does not fall up to the rock. This reasoning might be considered a final (or best) cause to explain why the rock falls. On the other hand, there is one additional step that the philosopher might take, which is to attempt to explain the "True Cause" (*vera causa.*) In this case, the philosopher might say that the True Cause of the rock falling is that a benevolent god designed the world in such a way that the law of gravity will always work consistently. This particular *teleological* explanation relies on a supernatural intervention and is the basis of natural theology.

Much of the furor surrounding Darwin's hypothesis in *The Origin of Species* resulted from his explanation of a *vera causa* using random variation and natural selection, rather than divine intent. Your role as a member of the Council of the Royal Society is to evaluate his work, but also to consider the implications of this kind of explanation of causality.

Two Methods in Nineteenth-Century Science: The Dispute over Induction and Hypothetical-Speculation

As a man of science in 1864, you are probably an advocate of the inductive method, unless your role sheet states otherwise. As a proponent of induction, you conform to a long string of philosophers stretching back to Francis Bacon. Bacon propounded a *modified* form of induction; this was not a simple collection of particulars to support a generalization. Rather, Bacon thought that investigators must actively manipulate the natural world to determine the causes of certain events; what we call experimentation.

For Bacon, when investigating the natural laws of the world, one first must identify a condition that precedes an event. For example, a modification to the movement of molecules is a condition that may lead to heat. The inquirer searches for positive instances of the connection: what variations in the initial condition (the modification to the movement of molecules) will still lead to the event (heat)? Then, the inquirer searches for *negative* instances: what variations in the initial condition will *not* lead to the event? Thus, the inquirer observes that an increase of heat is always preceded (in all observed cases) by an increase in the movement of particles. So, it may be the case that an increase in the movement of particles is the cause of heat; however, we cannot be certain. Although in all cases of heat there are cases of increased particle movement, we cannot conclude the opposite: that in all cases of increased particle movement, there is an increase in heat. Now the inquirer moves his attention from heat to increased particle movement and actively seeks instances where the increased movement is *not* followed by an increase in heat; these cases would be the *negative* instances. Once thorough research has been conducted and negative instances are not found, we can then conclude a natural law: when molecules rapidly *increase* their movement, then heat is always produced.

Following Bacon's method, Isaac Newton (1642–1727) discovered some of the most fundamental laws of nature. Newton adhered to Bacon's principles so perfectly that he became idolized for generations as the greatest man of science ever. The British poet, Alexander Pope (1688–1744), wrote "Nature and Nature's laws lay hid in night: God said, 'Let Newton be!' and all was light." His vast body of observation and experimentation, showing cases where laws do and do not apply, became the standard by which all other scientific theories were to be judged. Indeed, this was one of the criticisms that confronted Darwin: his method did not match Newton's.

The method of hypothetical-speculation contrasts with the method of induction because, whereas Bacon's method moves from the observation and experimentation on particular cases, slowly working up to induce more general statements explaining whole groups of phenomena, the hypothetical method creates an explanation through speculation rather than experimentation. Then, after the hypothesis is created, the pertinent events in the world are observed and tested to see how well the hypothesis persists. To clarify, consider the natural philosopher as someone who is exposed to a world filled with various phenomena of marvelous diversity, but, despite this diversity, the philosopher recognizes that there are some definite patterns at work in the world and believes that these patterns result from natural laws. This philosopher seeks "explanations" of the world by asking, "Why is the world the way that it is?" The philosopher begins with the basic assumption that all things in the world have a reason for being; nothing simply occurs without something else determining how it occurs. Now, how does the philosopher go about making sense of this plethora of natural phenomena?

To begin with, the philosopher believes two things about the world in which he resides. First, there are a multitude of individual objects and events, from stones and fish to magnetism and electricity. He considers each thing as distinct: each falling rock is distinct from other falling rocks and knowledge of falling rock 'A' is, therefore, distinct from knowledge of falling rock 'B.' Second, the philosopher recognizes patterns and believes that these patterns are the result of "natural laws" that govern how all things behave. Hence, knowledge of natural law 'X' provides knowledge of how *all things* behave.

As with the marbles, once he has observed one thousand falling rocks that all fall in a similar manner, he *induces* that if he had knowledge of some currently unknown falling rock, then that knowledge would be similar to that of the knowledge he currently possesses of known falling rocks. Therefore, if he can discover a consistent pattern in the falling of all known rocks, then he can induce that the pattern applies for other cases of falling rocks.

Although this method works well when inquiring into the nature of observable and testable matter, our natural philosopher thinks that there might be phenomena beyond the limits of his observation that are nevertheless worth knowing about. Certainly, this inductive method can still shed some light on things that *cannot possibly be observed*, in much the same way *that it can shed light on those falling rocks that could be, but have not been observed.* Occasionally, a set of phenomena is observed for which no explanation can be found; yet, an explanation is sought by the scientists. Natural philosophers search for a true cause that would explain this phenomenon. In order for this search to prove fruitful, the true cause must be of a certain type: it must be observable.

Given that our senses are limited and we are not omniscient, there may be phenomena that are impossible to observe. When this is the case, the only way to construct theories that explain the phenomena is to speculate, by creating a hypothesis, on what the true cause may be. The natural philosopher is thus employing induction,

Charles Darwin, the Copley Medal, and the Rise of Naturalism

but not as strongly as would be done in the inductive method. Rather than making a very small step from observable particulars to well-supported generalizations, the speculator makes a much bigger jump, bypassing well-supported generalizations to less supported ones. Once a law has been established through speculation, the basis for any explanation is essentially a figment, from which point a new interpretation on observed cases is provided. The theory then gives one the conceptual framework to *deduce* how the world must work in a particular aspect, without having to search for further observable causes. This conclusion is reflected in Sedgwick's analysis of Darwin's theory and the reason he said, "I look on the theory as a vast pyramid resting on its apex, and that apex a mathematical point." The apex is the theory of natural selection and Sedgwick is accusing Darwin of flying from the particulars to a general theory prematurely and arriving at his thesis through speculation.

Newton was a strong advocate of the inductive method over the hypothetical method. Within Newton's own philosophy, however, the controversy between the two methods persisted. Newton stated that his laws of motion and his theory of gravity depended on the absolute nature of time and space. Time moves at a constant rate, from past to present, at all places simultaneously, and space can be conceived on a perfect three dimensional grid. Both of these exist unchanged even when the objects within time do not undergo change and when space is devoid of anything. Newton's dependence and belief on the absolute nature of time and space, however, is itself a hypothesis. He never observed or tested time and space nor could he: they are abstract concepts. He simply assumed that they must exist in order for his laws to work. Thus, even the great Newton depended on the hypothesis-speculation method. It was only by accepting this hypothesis that natural philosophers could make real and practical advances in science and engineering, so his detractors were quickly silenced. (Newton's assumption turned out to be a major flaw and, in the wake of Ernest Mach and Albert Einstein, seventeenth-century notions about the absolute nature of time and space have been abandoned, marking the modern transition from Newtonian to quantum physics.)

Perspectives on Darwin's Method: A Critical Controversy

As one of the first assaults lobbed against Darwin's theory, Sedgwick's response became paradigmatic of the concerns surrounded the hypothetical nature of his work. John Herschel and John Stuart Mill (1806–1873), a British empiricist and social philosopher, both of whose work you know well in your role as a member of the Royal Society, were two major thinkers that dealt with scientific methodology in Darwin's era.

Herschel was a proponent of induction, accepted the hypothetical wave theory of light, and ridiculed the theory of natural selection. The technical reason that Herschel objected to natural selection was connected to his belief in "intelligent direction" as laid out in *The Physical Geography of the Globe*. In Herschel's view, Darwin was proposing a system where variation could happen in any random direction, not in a predetermined direction designed to help the organisms become more adaptable. This process of random variation leading to forms such as homo-sapiens seemed as incredible as supposing that a hundred monkeys could write a Shakespeare play. Herschel wrote, "Intelligence, guided by purpose, must be continually in action to bias the directions of the steps of change—to regulate their amount—to limit their divergence—and to continue them in a definite course" (Edinburgh: Adam and Charles Black, 1861, p. 12).

In Herschel's estimation, a purposeful God served as the primary cause of everything: God manipulates the universe through secondary causes that we know as the laws of nature. The natural philosopher attempted to discover the nature of those secondary laws and the patterns revealed there, which, in turn, aided the understanding of the revealed will of God, constituting the essence of the design argument. What the natural philosopher could not understand was the ultimate purpose of the primary cause (God). By arguing that natural selection operated exclusively through environmental conditions on random variation, Darwin affirmed that God was *not* the source of secondary causes or, at least, that the secondary causes could have existed without God. In other words, the secondary causes were a sufficient explanation in and of themselves and since selection was random, it did not reveal or depend on a purposeful design and designer. If God was not a purposeful designer of the universe, then what was He? Herschel puzzled over the implications of Darwin's theory in the passage below.

> We do not believe that Mr. Darwin means to deny the necessity of such intelligent direction. But it does not, so far as we can see, enter into the formula of his law; and without it we are unable to conceive how the law can have led to the results. On the other hand, we do not mean to deny that such intelligence may act according to a law (that is to say, on a preconceived and definite plan). Such law, stated in words, would be no other that the actual observed law of organic succession; or more general, taking that form when applied to our planet, and including all the links of the chain which have appeared. But the one law is a necessary supplement to the other, and ought, in all logical propriety, to form a part of its enunciation (*Physical Geography*, p.12).

John Stuart Mill was one of the few natural philosophers of the time who valued the method of hypothetical-speculation. For Mill, providing a hypothesis was the first major step in providing a scientific theory. The virtue of a good hypothesis is that it is set out in ways that can be tested; whether or not it is true is irrelevant in judging the value of a hypothesis. For example, in the eighteenth century, there was a theory that every disease originated in a particular organ of the body. That theory had already been proven wrong, but Mill thought that this was a good and worthwhile hypothesis because it was constructed in such a way that it could be tested. Concerning Darwin, he wrote in *A System of Logic*:

> Mr. Darwin's remarkable speculation on the Origin of Species is another unimpeachable example of a legitimate hypothesis. What he terms 'natural selection' is not only a *vera causa*, but one proved to be capable of producing effects of the same kind with those which the hypothesis ascribes to it: the question of possibility is entirely one of degree. It is unreasonable to accuse Mr. Darwin (as has been done) of violating the rules of Induction. The rules of induction are concerned with the conditions of proof. Mr. Darwin has never pretended that his doctrine was proved. He was not bound by the rules of Induction, but by those of Hypothesis. And these last have seldom been more completely fulfilled. He has opened a path of enquiry full of promise, the results of which none can foresee. And is it not a wonderful feat of scientific knowledge and ingenuity to have rendered so bold a suggestion, which the first impulse of every one was to reject at once, admissible and discussible, even as a conjecture? (*A System of Logic, Ratiocinative and Inductive, Being a Connected View of the Principles of Evidence, and the Methods of Scientific Investigation*, 7th ed., vol. 2, London: Longmans, Green Reader and Dryer, 1868, p. 18).

Mill supports Darwin's theory, but notice that he remains conditional. He agrees with the opponents of Darwin that he did *not* use induction, and thus that natural selection is not inductive, but hypothetical-speculative. According to the majority opinion of the day, this conclusion confirmed the argument that Darwin's theory was not scientific. Mill disagreed. In his view, Darwin's work stood as a testament to the legitimate role of hypothesis.

Conclusion

This section provides you with an introduction to the variegated ways that natural philosophers thought about science in the early 1860s. Although this summary is just the tip of the iceberg, use it as a starting point and re-read *On the Origin of Species* and the various reviews included in this packet and raise questions to yourself and others about how Darwin conducted his science and made his argument. When did Darwin use the inductive method? When did he use the hypothesis-speculative method? Which methodology was he employing? Are his critics wrong? Are there *verae causae*? Consider these questions, answer them carefully, and your arguments will certainly be improved.

THE HISTORICAL CONTEXT: THINGS YOU SHOULD KNOW

Religion in Great Britain

Religious history in Great Britain is the product of a fundamental tension—the struggle to create a unified national church versus the great diversity of religious belief throughout the country—and its corollary, volunteerism versus state coercion. Although the Church of England (the Anglicans) dominated the cultural landscape of the Victorian era, it had not always been so and the Anglicans had plenty of competition by the nineteenth century.

When Martin Luther presented his critique of the European church in 1517, others joined his effort to reform institutional Christianity. This movement became known as the Reformation and resulted in a split between those who remained loyal to established church authorities (the Catholics) and those who began to create a variety of new approaches to Christianity (the Protestants). Throughout Europe, these two versions of Christianity competed for political endorsements, as it was widely believed that government ought to support and specify the "true" church. Dissent from the official church was viewed as disloyalty to the state and condemned as a form of treason.

At first, the Tudors, who ruled England, remained loyal to Catholicism, but when, in the 1530s, Henry VIII failed to obtain from the Pope a writ of divorce from his wife, who had not produced a male heir, the English Crown began its own reformation. Because the king was motivated primarily by political considerations, actual changes in the English church focused on replacing the Pope with the crown as the supreme authority. Theology, organization, and even most of the ritual changed far less than in other Protestant countries. Henry VIII had two daughters, both raised as Catholics, and one son, Edward VI. Edward was a Protestant but died after only a few years as a very youthful king. English Catholics waited him out and happily celebrated the ascent of Henry's daughter Mary, who brought England back into the Catholic fold in 1553.

After Mary, too, died without producing an heir, Henry's second daughter, Elizabeth became queen in 1558. Though raised a Catholic, she preferred not to trust the Catholic supporters of (in her view) her treacherous sister and she converted to Protestantism. Wavering back and forth had been traumatic for the nation, creating a breeding ground for intrigue and disloyalty that sometimes necessitated executions. After re-establishing the Church of England, the new queen fashioned what became known as the "Elizabethan Settlement." She instituted a "don't ask, don't tell" rule for Catholics and assumed they were loyal to her as long as they worshiped privately and supported her politically. The Church of England, under Elizabeth's rule, struck a balance between Protestant theology and Catholic ritual, but vilified the Pope at every turn. Her solution is evident even today: reflected in Anglican ritual and a church hierarchy that retains traditionally labeled Catholic offices, such as bishop and archbishop.

Elizabeth's long reign was a comparatively peaceful period in church history, but she also died without an heir, leaving England vulnerable to instability. In 1603, England turned to her Stuart cousins who, while nominally Protestant, had Catholic sensibilities. In contrast, a large faction of both laity and clergy, as committed Protestants, wished to reform the Anglican Church more thoroughly and feared the influence of the Stuarts would take them in the wrong direction. These reformers became known as the Puritans and, eventually, they provoked a conflict with the second Stuart king (Charles I) in 1642: the English Civil War. Though the Puritans won the war and beheaded the king, in 1660, the Stuart line returned to the throne, raising suspicions about rising Catholic influence, once again. James II (the fourth Stuart king) was overthrown in a bloodless affair known as the Glorious Revolution of 1688. With the installation of William and Mary that year, England became firmly Protestant. The sixteenth-century documents *The Book of Common Prayer* and the Thirty-Nine Articles of Religion outlined its policies (moderately hierarchical), rituals (mildly Catholic), and theology (loosely Calvinist).

In 1690, Parliament passed an act of religious toleration in recognition of the fact that, though England was officially Protestant, the nation had to find a way to live in peace with the religious diversity that actually existed. As Great Britain incorporated Scotland and Ireland into the national entity, the government had to concede that the new Great Britain was not a "one size fits all" nation. Catholicism survived in England despite repeated attempts to eradicate it and there were numerous sects among the Protestants as well. Quakers, Methodists, and Presbyterians all attracted significant numbers in the eighteenth century and men of influence could be found in all religious faiths. In contrast to past history where religion and class were closely intertwined and the Church limited the prospects of those with the "wrong" religion, in the early nineteenth century, diverse religious perspectives were treated more tolerantly. Unitarians, in particular, took up the cause for toleration, extending their argument to include Jews as well as Christians of all types.

Despite these developments, toleration should not be confused with equality. In 1800, those who worshiped outside the Church of England were still considered to be dissenters and were denied many rights and privileges, including the right to hold public office and the privilege of attending major universities. Anglican clergy wielded more cultural authority than any other single group in the realm. Theirs was the final word on morality, ethics, truth, aesthetics, and most other cultural issues. They influenced politics and dominated education. The Church built a huge and effective bureaucracy that pervaded all areas of a citizen's life. You could live a comfortable life as a dissenter or Catholic, but you were well aware of your second-class status if you did not subscribe to the Thirty-Nine Articles.

By the mid-nineteenth century, several factors combined to shake the stronghold of the Anglican establishment: from foreign intellectual pressures, internal strife, and non-Anglican religious revivals that sprang up across Great Britain. *Wellhausen*, the new biblical criticism advocated by German academics and employed to great effect by some Anglican clergy, called for rethinking the Bible as a variety of literature, not unlike secular writing, created in a specific historical context and subject to scientific scrutiny. *Wellhausen* had a ripple effect across the Western world as holy men and laity began questioning the received wisdom about the Bible and God.

The Church also reeled under a series of scandals surrounding corrupt clergy, the Church's ineffectiveness in addressing social issues like poverty, and the lack of space to serve its national constituency. Millions simply chose not to attend Sunday services. Adding to these internal struggles was pressure from outside, as Revivalism attacked from a different direction, feeding the growth of evangelical sects like the Baptists and the Methodists. Street corner preachers were common in most urban areas as religious enthusiasts competed for converts, and a flood of revivalist religious publications bombarded the reading public.

Catholicism experienced a renewal, fueling dissent within the Church of England. The Oxford Movement (also known as the Tractarian Movement) of the 1830s emphasized the apostolic succession of Anglican clergy, veering dangerously close to Catholic doctrine. John Henry Newman, an important Anglican vicar who later converted to Catholicism, advocated a return to the church of the fourth century—what he termed an Anglo-Catholic faith. Increasingly viewed by the church establishment as a dangerous influence, Newman threatened to wreck the authority of the Thirty-Nine Articles as a means of distinguishing between the Catholic and Anglican faiths.

A number of political reforms encouraged more religious diversification. The Test and Corporation Acts, which required Anglican affiliation to hold public office, were rescinded in 1828. For the first time since the Union of 1800 had abolished the Dublin Parliament, Irish Catholics could sit in the British Parliament. Then again, although marriage in a dissenting church was legalized after 1836, for example, thorough reform took decades. As late as 1871, Oxford and Cambridge required its graduates to pledge an oath to the Anglican communion, even as denominational and dissenter universities were established. Nevertheless, at mid-century almost as many citizens attended other denominational services as worshiped in the official state-supported church.

By the time Darwin's ideas became the subject of public debate, there had already been considerable ferment within the Anglican Church. Even given the tenuous hold they had on their authority, it would be a mistake to think that Anglican authorities simply rejected Darwin's theories. Instead, there was a wide variety of responses and many clerics chose to embrace the new science.

Professional Organizations and Societies

The educated elite organized intellectual clubs and societies beginning in the seventeenth century. They exemplified the growing influence of education and science during the Enlightenment. By the Victorian era, they were a common feature of the social and political landscape. Association among well-educated men was a way of furthering common interests, popularizing the principles of science, pursuing healthy debate, and networking among the influential. Members usually met monthly or quarterly to read and discuss papers that outlined their most recent research and theories. While London was home to more clubs than any other city, many smaller urban centers also played host to regional associations. By Darwin's time, such groups saturated British society.

Organizations that focused primarily on science were based partly on the model offered by older literary and philosophical societies whose members met to discuss theology, philosophy, and, sometimes, science. The Anglican clergy, apt to be among the best educated and most influential men in their communities, and indeed the nation, were represented in large numbers because they had the means and time to engage in research and philosophical reflection and often counted a professorship amongst their accolades. Members of the gentry and landed aristocracy were also present in large numbers for similar reasons.

These new scientific organizations were also based on another model common in the nineteenth century: the reform society. Members of the British and Foreign Temperance Society or the British Anti-Slavery Society, for example, united to pursue specific political agenda and rally public support. Social rank and education were less important than unity around a common cause in these organizations.

By the mid-nineteenth century, scientific groups were increasingly organized along disciplinary lines. The Geological Society of London wielded considerable influence and the Royal Astronomical Society was also important. Others were much broader in conception, welcoming members in an array of fields. Both amateurs and academic professionals participated, although by the 1850s and 1860s professional and full-time scientists began to dominate. Two of the most important comprehensive groups were the British Association for the Advancement of Science (BAAS) and the more exclusive Royal Society (RS). Almost all of the characters in this game would have belonged to a number of societies.

Other scientific groups included the Royal Geographical Society, the Royal Society of Edinburgh, the Chemical Society, the Linnean Society, and the Statistical

Society. There was, in short, an organization for almost any branch of science, and their overlapping memberships formed a web of association that meant most men who engaged in scientific pursuits knew one another or at least could find acquaintances in common. British scientists navigated a small world.

Many men of social rank were members of the BAAS and non-members would often attend their meetings and demonstrations. Members of BAAS (founded in York in 1831) felt it was important to promote science as a source of legitimate cultural authority and held huge week-long meetings and scientific demonstrations annually. Regional urban centers competed to host the BAAS science rallies. Dramatic public demonstrations took on the aura of a circus, complete with bands, flags, and other entertainment. Because of its high profile, BAAS was sometimes lampooned in popular publications like *Punch* magazine, but BAAS also solicited serious scientific papers to be given at meetings and so combined the twin goals of furthering research and popularizing the results.

The Royal Society, by contrast, was much older (granted a royal charter in 1662) and staid in its approach. It occupied permanent rooms, provided by the crown, in London. Shortly after its founding, members (called Fellows and often referred to simply as FRS) began publishing *The Philosophical Transactions*, one of the most respected scientific journals in the world. By the nineteenth century, the RS also distributed money to further basic research and fund scientific expeditions. To become a member, an applicant had to be endorsed or sponsored by current a FRS. Aspiring Fellows who were controversial or had weak support risked being black-balled. Good social connections and appropriate rank greased the track and wealth helped. By 1840 there were about 750 FRS.

Beginning in the 1820s, a number of members, led by Charles Babbage, tried to reform the RS, feeling that it had grown increasingly moribund and that the organization contributed to a decline in scientific advancement in Great Britain. Signaling a shift toward increasing professionalization, after 1847 election as an FRS depended, at least in theory, solely on significant scientific achievement. As a result, membership numbers shrank steadily during the following decades. The fight over recognizing Charles Darwin's achievements constituted another chapter in the struggle to reform the RS.

In 1709, Sir Geoffrey Copley donated money to the RS to fund experiments. In 1736, Fellows proposed using the money for a medal to be awarded as an honorary prize for scientific achievement. By the time Darwin was nominated, the Copley Award's prestige was similar to receiving a Nobel Prize today.

The Enlightenment and Scottish Common Sense Philosophy

With the publication of Newton's *Principia* and John Locke's *Treatise on Human Affections* in the late seventeenth century, the Renaissance in Europe moved into its final phase, an intensely scientific and rationalist period referred to as the Enlightenment. The new emphasis on reason and belief in an orderly, knowable, and, therefore, predictable universe dominated thinking among the educated elite for over one hundred years. During this period, significant gains were made in scientific research, particularly regarding the physical and biological world, conducted primarily by independently wealthy amateurs, clergymen, and university men.

During its height, however, the Enlightenment saw little conflict between religion and science. Instead, most people who were interested in science also utilized what was known as the "design argument" or the "argument from design." This concept was often explained through the watchmaker analogy. If you found a watch (what they had in mind was a winding pocket watch with gears), but had no idea what it was, you might open the back to examine it. Inside you would find a set of tiny wheels, gears, and springs that fit together miraculously to turn the hands on the dial. Having seen how it worked, surely you would assume that not only did this device have a purpose; it must also have had an intelligent designer who built it with a purpose clearly in mind. By analogy, when we contemplate the complexities of the universe or the wondrous design of the human eye, surely we must conclude that there is purpose embedded by an intelligent creator in those designs. Enlightenment writers everywhere embraced the design argument as a convenient way to mediate between science and religion. William Paley wove it into the text of *Natural Theology; or Evidences of the Existence and Attributes of the Deity* (1802), excerpted in Appendix B.

In the wake of the violence and turmoil of the French Revolution, beginning in 1789 and wreaking havoc on the stability of Europe for the next several decades, many European intellectuals began to question what had become the comfortable truths of the Age of Reason. A number of ideologies (a term first coined during the French Revolution) emerged from the confusion of war and destruction, including liberalism (characterized by moderation and support for Enlightenment ideals with an emphasis on reform, civil liberties, and economic growth through free markets) and socialism (characterized by a desire to pursue social justice for those outside the middle class through a more dramatic reordering of society as a whole.)

As the nineteenth century progressed, a new philosophical and aesthetic outlook known as Romanticism swept over the continent. Whereas Enlightenment philosophy emphasized the use of reason and the idea that we understand the universe through scientific analysis, Romantics contended that science, useful for study of the physical world, could not resolve critical issues such as the reality of God or the purpose of human existence. They emphasized intuition over reason, thought of the world in organic rather than mechanical terms, and valued chaos and conflict as viable alternatives to order and harmony. As they elevated the importance of the individual role in reshaping the rigid class structures and outdated autocracy of monarchical governments, revolution became their rallying cry.

Throughout the nineteenth century, Romanticism competed with the persistent influence of the Enlightenment in European thought. One stream of modified Enlightenment ideas that continued in Great Britain (and was dominant in the United States) became known as Scottish Common Sense philosophy, most cogently expressed in the writings of Thomas Reid. Reid argued for the common sense of ordinary evidence and the rational judgment of ordinary people. Along with David Hume and Adam Smith, Reid emphasized sentiment and intuition as sources of belief. Everyone has beliefs that grow out of what we might think of as gut feelings, beliefs that have no real rational basis and cannot be proved. Reid heartily approved of such beliefs and classed the belief in God as an example of a proposition that is impossible to prove, but nonetheless true. He did not intend, however, to jettison key principles of the Enlightenment. Rather, the tenor of his argument is that moral judgment is the result of rational thought that produces feelings on which we act. Many scientists on both sides of the Atlantic seized upon Common Sense principles as a way to mediate between their scientific work and their religious beliefs.

Bridgewater Treatises

In 1829, the Earl of Bridgewater died, leaving a fund to the RS for publication of a tract he called, "On the Power, Wisdom, and Goodness of God, as Manifested in the Creation . . ." In consultation with the Archbishop of Canterbury and the Bishop of London, the President of the RS selected eight authors (among them the Reverend William Whewell) to pen a series of books, published between 1833 and 1836, that became known as the *Bridgewater Treatises*. The specific topics ranged from chemistry to biology to astronomy; all the authors were respected men of science and, with one exception, were FRS. Because the treatises were sold separately, few private libraries likely contained the entire set.

Bridgewater intended that the works would extend and support the argument of Paley's *Natural Theology* (1802). These efforts added up to a celebration and buttressing of the design argument. Primarily descriptive in approach, the *Treatises* tend to pile fact upon fact and represented the orthodoxy espoused by the Anglican hierarchy. Central to their argument was the assertion that God and nature might be poorly understood by man, and the Almighty might be a distant entity, but, in the end, there was no conflict that pitted his natural creation against his own being.

For the most part, however, the *Bridgewater Treatises* carefully separate religious and scientific pursuits, and, in that sense, they moved toward a less religious interpretation of natural phenomena because, after claiming that God is the divine author, they do not use religious reasons (e.g. angels) to explain the inner-workings of scientific nature. Though they often resorted to utilitarianism, in other instances the Treatises simply resorted to the mystery of a distant God for an explanation. In either case, they firmly endorsed the superiority and stewardship of human beings over the flora and fauna of God's creation.

Darwin was keenly aware of Paley and the *Bridgewater Treatises* as he wrote and thought about evolution, although his most explicit public denunciations of their assumptions came late, appearing in his *The Descent of Man* (1871) rather than in *On the Origin of Species* (1859). As Darwin and others in the scientific community came to realize, Natural Theology had to fall by the wayside in order for a more naturalistic science to gain the upper hand in popular culture.

Essays and Reviews

Published in an American edition in Boston by Frederic H. Hedge in 1860, *Essays and Reviews* introduced higher criticism of the Bible (originated in the German academy) into the whirlwind of controversy over Darwin's *On the Origin of Species*. The volume gained additional notoriety because all seven authors were ordained priests in the Church of England. In essence, they argued that biblical narrative should be compared to other historical sources to check for accuracy, and that, like any other book, the Bible should be subject to criticism and contextualization. Such propositions required readers of the Bible to consider that its authors lived in a particular time and place, came to their writing with the prejudices of any other authors, and wrote from particular points of view. The overall implication was immediately apparent: the Bible was not written by God's hand.

Many geologists and biologists welcomed *Essays and Reviews* as compatible with their work. Others both inside and outside of scientific circles condemned its authors as "The Seven Against Christ." Benjamin Jowett's "On the Interpretation of Scripture" remains a key essay in the book and merits the attention of all players. *Essays and Reviews* helped clarify what Darwin knew only too well: men of science could no longer avoid contemplating the idea that truth could be a matter of rational inquiry, not subject to the unexamined assumptions of Christian belief. In other words, the Bible might not represent a bedrock of truth (the revealed will of God) against which all other truths should be measured.

Appendices

APPENDIX A. DARWIN, *ON THE ORIGIN OF SPECIES* (1859)

Charles Darwin's *On the Origins of Species* is unquestionably one of the most culturally transformative achievements in human intellectual history. The edited selections below will introduce you to Darwin's approach to science as well as his fundamental arguments. Although students are encouraged to read the unabridged *On the Origin of Species*, much of the science is outmoded and unnecessary to understanding its theoretical basis. We recommend the edited version to increase the focus on the issues debated in this game. Janet Brown's *Darwin's "Origin of Species"* (2007) is a highly recommended supplemental text. You should also read the primary source documents available in Appendix B.

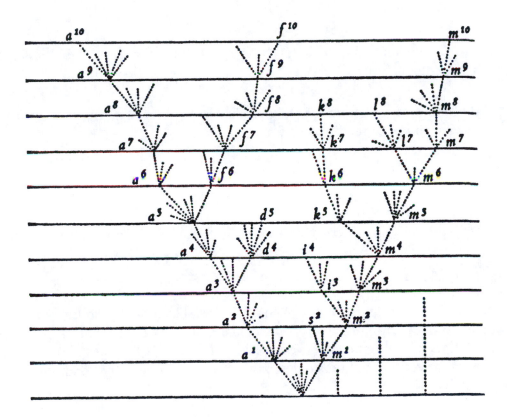

Figure 1. Lone figure from *On the Origin of Species*. (From a facsimile edition of *On the Origin of Species* by Charles Darwin (1872). Courtesy of the Library of Congress.

This annotated abridgement is intended to make Darwin's ideas more accessible to a broader audience and to provide historical context and connections to current evolutionary thinking. This is an evolving "social text" and suggestions or criticisms are welcome and should be directed to Dann Siems, Department of Biology, Bemidji State University, Bemidji MN 56601 – dsiems@bemidjistate.edu.

A SUMMARY OF THE LOGIC OF NATURAL SELECTION

Three inductions from observation and two necessary deductive conclusion

If Inductions 1-3 (derived directly from observation) are true then the Deductions (A & B) necessarily follow. The inductions derived from observations are corroborated by empirical evidence therefore evolution by natural selection is inevitable; all that remains as controversial it to determine its relative significance as a causal process.

Induction 1 – Adults on average produce (many) more offspring than required for their own replacement

Induction 2 – Populations remain relatively constant in number (at least they don't increase continuously)

> **Deduction A** – Therefore, it follows that some (many) offspring must fail to survive and/or to reproduce.

Induction 3 – Within any population there are heritable variations in form and physiology (ie., species have no immutable essence)

> **Deduction B** – Any heritable variations which enhance prospects for survival and reproduction will increase in frequency over time within a population (this is Natural Selection)

IMPORTANT NOTE: Deduction A in no way implies the inevitability of competition. Most offspring fall prey to predators, are victims of pathogens or parasites, or are victims of environmental events. The widespread belief (past and present!) that natural selection requires competition reflects cultural rather than biological foundations!

Charles Darwin, the Copley Medal, and the Rise of Naturalism

TABLE I. STRUCTURE AND ORIGINS OF DARWIN'S STRATEGY OF ARGUMENT(S) IN *ORIGIN OF SPECIES*

Part	Strategy	Tactic	Chapters
I. Variation and selection under domestication	*Vera causa* existence [after Herschel]	Establish accepted idea	*I. Variation under domestication*
		Argue from analogy	*II. Variation under nature* *III. Struggle for existence*
II. Variation and selection in nature	*Vera causa* competence	Make case	*IV. Natural selection* *V. Laws of variation*
		Consider difficulties	*VI. Difficulties of the theory* *VII*. Miscellaneous objections (added in 2nd edition)* *VII. Instinct* *VIII. Hybridism* *IX. Imperfections of the geological record*
III. Explanatory trials of the theory	*Vera causa* responsibility	Present evidence favoring responsibility	*X. Geological succession* *XI. Geographic distribution*
	Consilience of inductions [after Whewell]	Make sense of a large class of otherwise disparate facts	*XII. Geographic distribution* *XIII. Mutual affinities of organic beings*
Recapitulation	Allay fears, Convert ready	Humility & Reverie [after Humboldt]	*XIV. Recapitulation and conclusions*

Darwin built his case in *On the Origin of Species* specifically to address Victorian expectations concerning the nature of sound scientific practices. In particular, Darwin presents a *vera causa* ("true cause") argument in Chapters 1–11 that is a direct response to the practices advocated by John Herschel in "Preliminary Discourse on the Study of Natural Philosophy" (1830).

In a *vera causa* approach, one must first demonstrate the <u>existence</u> of some potentially causal process, then demonstrate that the process is in principle <u>competent</u> to explain the phenomenon of interest, and finally that the process is in fact <u>responsible</u>. In Chapters 12 and 13 Darwin follows the approach put forth by William Whewell, the first philosopher of science, in his "History (1837) and Philosophy (1840) of the Inductive Sciences." Whewell (who incidentally had coined the then rather controversial term 'scientist' in 1833) argued that the quality and utility of a scientific theory could be judged based on its capacity to makes sense of large class of otherwise apparently unrelated facts and coincidences. In carefully and intentionally structuring his argument in this way, Darwin was actively seeking the approval of Herschel and Whewell as a strategy for enhancing the likelihood of a positive scientific and public reaction to his theory of evolution by natural selection. In the final chapter, Darwin echoes the 'nature reveries' of Humboldt's

"Personal Narrative of Travels to Equinoctial Regions of the New Continent" which Darwin had read as a youth and which shaped his rather Romantic views of nature throughout his life. Table adapted primarily from Hodge, M. 1977. The Structure and strategy and Darwin's 'Long Argument.' *British Journal for the History of Science 10:* 237-246 and Waters, K. The arguments in the Origin of Species. pp. 116-140. In: Hodge, M. & Radick, G. *The Cambridge Companion to Darwin.* Cambridge University Press, 2003; see also Sloan, P.R. 1991. The sense of sublimity: Darwin on nature and the divine. Osiris 16: 251-269. Janet Browne's (2007) *Darwin's "Origin of Species": A Biography* provides a concise analysis of the context and development of Darwin's theory.

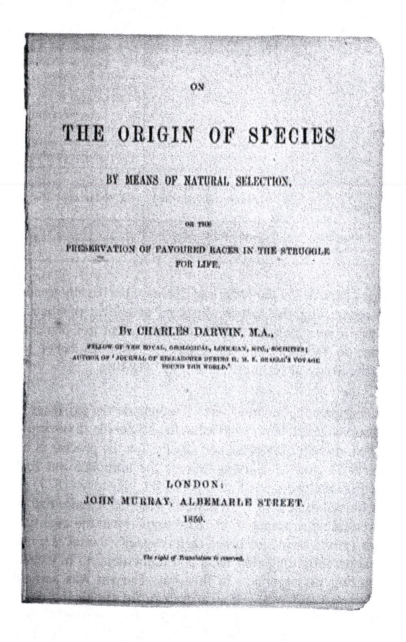

Figure 2. Title page from *On the Origin of Species* (From Wikimedia Commons)

Charles Darwin, the Copley Medal, and the Rise of Naturalism

ON THE ORIGIN OF SPECIES BY MEANS OF NATURAL SELECTION, OR THE PRESERVATION OF FAVOURED RACES IN THE STRUGGLE FOR LIFE.[1,2]

Charles Darwin, M.A.
Fellow of the Royal, Geological, Linnean, etc. societies;
Author of Journal of researches during H. M. S. Beagle's Voyage round the world.
London: John Murray, Albemarle Street, 1859
—Edited and annotated by Dann Siems (2007)[3]

PREFACE—AN HISTORICAL SKETCH

I WILL here give a brief sketch of the progress of opinion *On the Origin of Species*. Until recently the great majority of naturalists believed that species were immutable productions, and had been separately created. This view has been ably maintained by many authors. Some few naturalists, on the other hand, have believed that species undergo modification, and that the existing forms of life are the descendants by true generation of pre-existing forms. [. . .]

Lamarck was the first man whose conclusions on the subject excited much attention. This justly-celebrated naturalist first published his views in 1801; he much enlarged them in 1809 in his "Philosophie Zoologique," and subsequently, in 1815, in the Introduction to his "Hist. Nat. des Animaux sans Vertébres." In these works he upholds the doctrine that species, including man, are descended from other

---

1 The first edition of this work (1,250 copies) was published on November 24th, 1859. In seeking support for his theory, Darwin sent complimentary copies to one hundred leading men of science. A second edition (3,000 copies) following quickly on January 7th, 1860. A Preface first appeared in the third edition (2,000 copies) published in 1861 but the preface version reproduced here is from the 6th edition published in 1872 (3,000 copies; 4th edition 1,500 copies) Interestingly, Herbert Spencer's phrase "survival of the fittest" did not appear in the Origin until the 5th edition published in 1869 (2,000 copies). Spencer had coined this language in his 1864 book "Principles of Biology" where he wrote: "This survival of the fittest, which I have here sought to express in mechanical terms, is that which Mr. Darwin has called 'natural selection', or the preservation of favoured races in the struggle for life." Although many of Darwin's contemporaries urged him to use "survival of the fittest" as a synonym for natural selection in the Origin, Darwin consistently resisted these entreaties believing Spencer's phrase simplistic and misleading. [Note that only 12,750 copies of the Origin were printed in the first decade; however, it is likely that most, if not all members of the Royal Society had a copy or were familiar with its contents].

2 First edition full text is available online at http://www.talkorigins.org/faqs/origin.html. John van Wyhe provides another excellent on-line first edition at http://darwin-online.org.uk; this site has the advantage of preserving original page numbering.

3 Selections reproduced here are intended to provide an abridged overview of the style and structure of Darwin's argument. **Game players should pay particular attention to the introductory remarks at the head of each chapter as they highlight the conceptual content and the strategic rhetorical role of subsequent material. In addition, GAME NOTES specific to various issues are interspersed throughout the text as are CHARACTER ALERTS specific to particular roles.** Editorial comments and annotations are presented in [*italicized brackets*] or footnotes and include brief descriptions of deleted material, contextual commentary, and connections to contemporary evolutionary theory.

species. He first did the eminent service of arousing attention to the probability of all change in the organic, as well as in the inorganic world, being the result of law, and not of miraculous interposition. Lamarck seems to have been chiefly led to his conclusion on the gradual change of species, by the difficulty of distinguishing species and varieties, by the almost perfect gradation of forms in certain groups, and by the analogy of domestic productions. With respect to the means of modification, he attributed something to the direct action of the physical conditions of life, something to the crossing of already existing forms, and much to use and disuse, that is, to the effects of habit. To this latter agency he seemed to attribute all the beautiful adaptations in nature;—such as the long neck of the giraffe for browsing on the branches of trees. But he likewise believed in a law of progressive development; and as all the forms of life thus tend to progress, in order to account for the existence at the present day of simple productions, he maintains that such forms are now spontaneously generated.

[. . .] It is curious how largely my grandfather, Dr. Erasmus Darwin, anticipated the views and erroneous grounds of opinion of Lamarck in his "Zoonomia" [. . .] published in 1794. According to Isid. Geoffroy Saint Hilaire there is no doubt that Goethe was an extreme partisan of similar views, as shown in the Introduction to a work written in 1794 and 1795, but not published till long afterwards. [. . .] It is rather a singular instance of the manner in which similar views arise at about the same time, that Goethe in Germany, Dr. Darwin in England, and Geoffroy Saint-Hilaire [. . .] in France; came to the same conclusion *On the Origin of Species*, in the years 1794–5.

[There follows a compilation of thirty-four examples of others who apparently espoused evolutionary views and even some who recognized the principle of natural selection between 1800 and 1859. Interestingly, Darwin does not claim that he was the first to propose the idea of natural selection and he even uses prior recognition of the principle by others to support the legitimacy of his argument. Only several of the excerpts most relevant to present concerns are included here. **Game Note:** *Look for references to your character (or an ally or adversary) of your character here. Darwin's comments on Owen are particularly revealing.]*

The "Vestiges of Creation" appeared in 1844. In the tenth and much improved edition (1853) the anonymous author[4] says (p. 155):—"The proposition determined on after much consideration is, that the several series of animated beings, from the simplest and oldest up to the highest and most recent, are, under the providence of God, the results, first, of an impulse which has been imparted to the forms of life, advancing them, in definite times, by generation, through grades of organisation

---

4 Revealed to be Robert Chambers in 1884. This work was widely read on both sides of the Atlantic and brought natural history to public attention. For more information, see http://www.ucmp.berkeley.edu/history/chambers.html.

terminating in the highest dicotyledons and vertebrata, these grades being few in number, and generally marked by intervals of organic character, which we find to be a practical difficulty in ascertaining affinities; second, of another impulse connected with the vital forces, tending, in the course of generations, to modify organic structures in accordance with external circumstances, as food, the nature of the habitat, and the meteoric agencies, these being the 'adaptations' of the natural theologian." The author apparently believes that organisation progresses by sudden leaps, but that the effects produced by the conditions of life are gradual. He argues with much force on general grounds that species are not immutable productions. [. . .] The work, from its powerful and brilliant style, though displaying in the earlier editions little accurate knowledge and a great want of scientific caution, immediately had a very wide circulation. In my opinion it has done excellent service in this country in calling attention to the subject, in removing prejudice, and in thus preparing the ground for the reception of analogous views. [. . .]

Professor Owen, in 1849 ("Nature of Limbs," p. 86), wrote as follows:—"The archetypal idea was manifested in the flesh under diverse such modifications, upon this planet, long prior to the existence of those animal species that actually exemplify it. To what natural laws or secondary causes the orderly succession and progression of such organic phenomena may have been committed, we, as yet, are ignorant." In his Address to the British Association, in 1858, he speaks (p. li.) of "the axiom of the continuous operation of creative power, or of the ordained becoming of living things." Farther on (p. xc.), after referring to geographical distribution, he adds, "These phenomena shake our confidence in the conclusion that the Apteryx of New Zealand and the Red Grouse of England were distinct creations in and for those islands respectively. Always, also, it may be well to bear in mind that by the word 'creation' the zoologist means 'a process he knows not what.'" He amplifies this idea by adding that when such cases as that of the Red Grouse are enumerated by the zoologists as evidence of distinct creation of the bird in and for such islands, he chiefly expresses that he knows not how the Red Grouse came to be there, and there exclusively; signifying also, by this mode of expressing such ignorance, his belief that both the bird and the islands owed their origin to a great first Creative Cause. If we interpret these sentences given in the same Address, one by the other, it appears that this eminent philosopher felt in 1858 his confidence shaken that the Apteryx and the Red Grouse first appeared in their respective homes, "he knew not how," or by some process "he knew not what."

This Address was delivered after the papers by Mr. Wallace and myself *On the Origin of Species*, presently to be referred to, had been read before the Linnean Society. When the first edition of this work was published, I was so completely deceived, as were many others, by such expressions as "the continuous operation of creative power," that I included Professor Owen with other palaeontologists as being firmly convinced of the immutability of species; but it appears ("Anat. of Vertebrates," vol. iii. p. 796) that this was on my part a preposterous error. In the last edition of this work I inferred, and the inference still seems to me perfectly just, from a passage beginning with the words "no doubt the type-form," &c. (Ibid., vol. i. p. xxxv.), that Professor Owen admitted that natural selection may have done something in the formation of a new species; but this it appears (Ibid., vol. nl. p. 798) is inaccurate and without evidence. I also gave some extracts from a correspondence between Professor Owen and the Editor of the "London Review," from which it appeared manifest to the Editor as well as to myself, that Professor Owen claimed to have promulgated the theory of natural selection before I had done so; and I expressed my surprise and satisfaction at this announcement; but as far as it is possible to understand certain recently published passages (Ibid., vol. iii. p. 798) I have either partially or wholly again fallen into error. It is consolatory to me that others find Professor Owen's controversial writings as difficult to understand and to reconcile with each other, as I do. As far as the mere enunciation of the principle of natural selection is concerned, it is quite immaterial whether or not Professor Owen preceded me, for both of us, as shown in this historical sketch, were long ago preceded by Dr. Wells and Mr. Matthews. [. . .]

The "Philosophy of Creation" has been treated in a masterly manner by the Rev. Baden Powell, in his "Essays on the Unity of Worlds," 1855. Nothing can be more striking than the manner in which he shows that the introduction of new species is "a regular, not a casual phenomenon," or, as Sir John Herschel expresses it, "a natural in contradistinction to a miraculous, process."[5]

[. . .] Professor Huxley gave a lecture before the Royal Institution on the "Persistent Types of Animal Life." Referring to such cases, he remarks, "It is difficult to comprehend the meaning of such facts as these, if we suppose that each species of animal and plant, or each great type of organisation, was formed and placed upon the surface of the globe at long intervals by a distinct act of creative power; and it is well to recollect that such an assumption is as unsupported by tradition or revelation as it is opposed to the general analogy of nature. If, on the other hand, we view 'Persistent Types' in relation to that hypothesis which supposes the species living at any time to be the result of the gradual modification of pre-existing species a hypothesis which, though unproven, and sadly damaged by some of its supporters, is yet the only one to which physiology lends any countenance; their existence would seem to show that the amount of modification which living beings have undergone during geological time is but very small in relation to the whole series of changes which they have suffered."

. . . Dr. Hooker published his "Introduction to the Australian Flora." In the first part of this great work he admits the truth of the descent and modification of species, and supports this doctrine by many original observations.

---

5 Baden-Powell was also one of seven contributors to "Essay & Reviews" (1860) which precipitated the events that are central to the Darwin game. Incidentally, Baden-Powell's son later founded the Boy Scouts.

CONTENTS

INTRODUCTION

CHAPTER I

VARIATION UNDER DOMESTICATION

CHAPTER II

VARIATION UNDER NATURE

CHAPTER III

STRUGGLE FOR EXISTENCE

CHAPTER IV

NATURAL SELECTION

CHAPTER V

LAWS OF VARIATION

CHAPTER VI

DIFFICULTIES OF THE THEORY

CHAPTER VII

INSTINCT

CHAPTER VIII

HYBRIDISM

CHAPTER IX

ON THE IMPERFECTION OF THE GEOLOGICAL RECORD

CHAPTER X

ON THE GEOLOGICAL SUCCESSION OF ORGANIC BEINGS

CHAPTER XI

GEOGRAPHICAL DISTRIBUTION

CHAPTER XII

GEOGRAPHICAL DISTRIBUTION—continued

CHAPTER XIII

MUTUAL AFFINITIES OF ORGANIC BEINGS: MORPHOLOGY: EMBRYOLOGY: RUDIMENTARY ORGANS

CHAPTER XIV

RECAPITULATION AND CONCLUSIONS

GLOSSARY (34—FROM 6TH EDITION: AVAILABLE ONLINE)

INTRODUCTION

[Game Note: Before proceeding with the text itself, pay particularly close attention to the structure of Darwin's argument (see "Summary of the Logic of Natural Selection" and "Table I: Structure of Darwin's Argument" at the beginning of this appendix). Note especially that Darwin devotes four (or, in later editions, five) chapters to considering "the most apparent and gravest difficulties" facing his theory. This self-critical approach is an important forerunner to the now standard practice of placing one's own "pet hypotheses" at maximum jeopardy.]

WHEN on board H.M.S. Beagle, as naturalist, I was much struck with certain facts in the distribution of the inhabitants of South America, and in the geological relations of the present to the past inhabitants of that continent. These facts seemed to me to throw some light *On the Origin of Species*—that mystery of mysteries, as it has been called by one of our greatest philosophers. On my return home, it occurred to me, in 1837, that something might perhaps be made out on this question by patiently accumulating and reflecting on all sorts of facts which could possibly have any bearing on it. After five years' work I allowed myself to speculate on the subject, and drew up some short notes; these I enlarged in 1844 into a sketch of the conclusions, which then seemed to me probable: from that period to the present day I have steadily pursued the same object. I hope that I may be excused for entering on these personal details, as I give them to show that I have not been hasty in coming to a decision. *[Game Note: Why is Darwin so concerned not to be perceived as "hasty?"]*

My work is now nearly finished; but as it will take me two or three more years to complete it, and as my health is far from strong, I have been urged to publish this Abstract. I have more especially been induced to do this, as Mr. Wallace, who is now studying the natural history of the Malay archipelago, has arrived at almost exactly the same general conclusions that I have *On the Origin of Species*. Last year he sent to me a memoir on this subject, with a request that I would forward it to Sir Charles Lyell, who sent it to the Linnean Society, and it is published in the third volume of the journal of that Society. Sir C. Lyell and Dr. Hooker, who both knew of my work—the latter having read my sketch of 1844—honoured me by thinking it advisable to publish, with Mr. Wallace's excellent memoir, some brief extracts from my manuscripts.[6]

This Abstract, which I now publish, must necessarily be imperfect. I cannot here give references and authorities for my several statements; and I must trust to the

---

6 Papers were read 1 July 1858. Darwin's paper was read by George Busk, a member of the X-Club who later would nominate Darwin for the Royal Society's Copley Medal. Copies of both papers are available online at: http://www.darwingame.org/Darwin and Wallace 1858.pdf.

reader reposing some confidence in my accuracy. No doubt errors will have crept in, though I hope I have always been cautious in trusting to good authorities alone. I can here give only the general conclusions at which I have arrived, with a few facts in illustration, but which, I hope, in most cases will suffice. No one can feel more sensible than I do of the necessity of hereafter publishing in detail all the facts, with references, on which my conclusions have been grounded; and I hope in a future work to do this. For I am well aware that scarcely a single point is discussed in this volume on which facts cannot be adduced, often apparently leading to conclusions directly opposite to those at which I have arrived. A fair result can be obtained only by fully stating and balancing the facts and arguments on both sides of each question; and this cannot possibly be here done.

I much regret that want of space prevents my having the satisfaction of acknowledging the generous assistance which I have received from very many naturalists, some of them personally unknown to me. I cannot, however, let this opportunity pass without expressing my deep obligations to Dr. Hooker, who for the last fifteen years has aided me in every possible way by his large stores of knowledge and his excellent judgment.

In considering the Origin of Species, it is quite conceivable that a naturalist, reflecting on the mutual affinities of organic beings, on their embryological relations, their geographical distribution, geological succession, and other such facts, might come to the conclusion that each species had not been independently created, but had descended, like varieties, from other species. Nevertheless, such a conclusion, even if well founded, would be unsatisfactory, until it could be shown how the innumerable species inhabiting this world have been modified so as to acquire that perfection of structure and co-adaptation which most justly excites our admiration. Naturalists continually refer to external conditions, such as climate, food, &c., as the only possible cause of variation. In one very limited sense, as we shall hereafter see, this may be true; but it is preposterous to attribute to mere external conditions, the structure, for instance, of the woodpecker, with its feet, tail, beak, and tongue, so admirably adapted to catch insects under the bark of trees. In the case of the misseltoe, which draws its nourishment from certain trees, which has seeds that must be transported by certain birds, and which has flowers with separate sexes absolutely requiring the agency of certain insects to bring pollen from one flower to the other, it is equally preposterous to account for the structure of this parasite, with its relations to several distinct organic beings, by the effects of external conditions, or of habit, or of the volition of the plant itself.

The author of the "Vestiges of Creation" would, I presume, say that, after a certain unknown number of generations, some bird had given birth to a woodpecker, and some plant to the misseltoe, and that these had been produced perfect as we now see them; but this assumption seems to me to be no explanation, for it leaves the case of the coadaptations of organic beings to each other and to their physical conditions of life, untouched and unexplained.

It is, therefore, of the highest importance to gain a clear insight into the means of modification and coadaptation. At the commencement of my observations it seemed to me probable that a careful study of domesticated animals and of cultivated plants would offer the best chance of making out this obscure problem. Nor have I been disappointed; in this and in all other perplexing cases I have invariably found that our knowledge, imperfect though it be, of variation under domestication, afforded the best and safest clue. I may venture to express my conviction of the high value of such studies, although they have been very commonly neglected by naturalists.

From these considerations, I shall devote the first chapter of this Abstract to Variation under Domestication. We shall thus see that a large amount of hereditary modification is at least possible, and, what is equally or more important, we shall see how great is the power of man in accumulating by his Selection successive slight variations. I will then pass on to the variability of species in a state of nature; but I shall, unfortunately, be compelled to treat this subject far too briefly, as it can be treated properly only by giving long catalogues of facts. *[Game Note: Why does Darwin note the need for "long catalogues of facts"? Remember that Darwin intended the* Origin *as a brief abstract of his theory. If your character needs these additional examples you may find it helpful to refer to some of his other works (e.g., on barnacles or orchids). To understand more fully the importance of "catalogues of fact" refer to section on "Playing a Natural Philosopher."]* We shall, however, be enabled to discuss what circumstances are most favourable to variation. In the next chapter the Struggle for Existence amongst all organic beings throughout the world, which inevitably follows from their high geometrical powers of increase, will be treated of. This is the doctrine of Malthus, applied to the whole animal and vegetable kingdoms. As many more individuals of each species are born than can possibly survive; and as, consequently, there is a frequently recurring struggle for existence, it follows that any being, if it vary however slightly in any manner profitable to itself, under the complex and sometimes varying conditions of life, will have a better chance of surviving, and thus be naturally selected. From the strong principle of inheritance, any selected variety will tend to propagate its new and modified form. *[Game Note: Is a competitive "struggle for existence" necessarily a consequence of surplus reproduction? See additional comments in Chapter III notes.]*

This fundamental subject of Natural Selection will be treated at some length in the fourth chapter; and we shall then see how Natural Selection almost inevitably causes much Extinction of the less improved forms of life and induces what I have called Divergence of Character. In the next chapter I shall discuss the complex and little known laws of variation and of correlation of growth. In the four succeeding chapters, the most apparent and gravest difficulties on the theory will be given: namely, first, the difficulties of transitions, or understanding how a simple being or a simple organ can be changed and perfected into a highly developed being or elaborately constructed organ; secondly the subject of Instinct, or the mental powers of animals, thirdly, Hybridism, or the infertility of species and the fertility of varieties when intercrossed; and fourthly, the imperfection of the Geological Record. In the next chapter I shall consider the geological succession of organic beings throughout time; in the eleventh and twelfth, their geographical distribution throughout space; in the thirteenth, their classification or mutual affinities, both when mature and in an embryonic condition. In the last chapter I shall give a brief recapitulation of the whole work, and a few concluding remarks.)

No one ought to feel surprise at much remaining as yet unexplained in regard to the origin of species and varieties, if he makes due allowance for our profound ignorance in regard to the mutual relations of all the beings which live around us. Who can explain why one species ranges widely and is very numerous, and why another allied species has a narrow range and is rare? Yet these relations are of the highest importance, for they determine the present welfare, and, as I believe, the future success and modification of every inhabitant of this world. Still less do we know of the mutual relations of the innumerable inhabitants of the world during the many past geological epochs in its history. Although much remains obscure, and will long remain obscure, I can entertain no doubt, after the most deliberate study and dispassionate judgement of which I am capable, that the view which most naturalists entertain, and which I formerly entertained—namely, that each species has been independently created—is erroneous.[7] I am fully convinced that species are not immutable; but that those belonging to what are called the same genera are lineal descendants of some other and generally extinct species, in the same manner as the acknowledged varieties of any one species are the descendants of that species. Furthermore, I am convinced that Natural Selection has been the main but not exclusive means of modification.

*[**Game Note:** Darwin provided a convenient sketch of the topics considered in each chapter. I have used bold-faced type to indicate those sections from which excerpts have been drawn. Players may what to consult an online, full text version to review relevant sections not included here.]*

7 In an 1844 letter to J.D. Hooker in which he revealed his evolutionary ideas for the first time, Darwin wrote, "I am almost convinced (quite contrary to the opinion I started with) that species are not (it is like confessing a murder) immutable. Heaven forfend me from Lamarck nonsense of a 'tendency to progression' 'adaptations from the slow willing of animals' &c, but the conclusions I am led to are not widely different than his—though the means of change are wholly so—I think I have found out (here's presumption!) the simple way by which species become exquisitely adapted to various ends." See R. Colp. (1986) "Confessing a Murder" – Darwin's First Revelations about Transmutation. Isis 77: 8–32 for more details on this letter. This paper also provides a good introduction to Darwin's earliest thinking about transmutation and is particularly insightful with respect to Darwin's personal reluctance to press his views publically. This work also offers a revealing portrayal of the relationship between Charles and Emma and shows that while she was long aware of his ideas she remained more than a little skeptical.

Charles Darwin, the Copley Medal, and the Rise of Naturalism

CHAPTER I

VARIATION UNDER DOMESTICATION

Causes of Variability – Effects of Habit and the Use of Disuse of Parts – Correlation of Growth – **Inheritance** – Character of Domestic Varieties – **Difficulty of distinguishing between Varieties and Species** – Origin of Domestic Varieties from one or more Species – **Domestic pigeons, their Differences and Origin – Principle of Selection anciently followed, its Effects – Methodical and Unconscious Selection** – Unknown Origin of our Domestic Productions – Circumstances favourable to Man's power of Selection

*[**Game Note:** Darwin's strategy in Chapter I is to establish that considerable variation exists in domesticated populations and that this variation is not an anomaly but instead is an inevitable result of processes of heredity and development. The existence and extent of this variation within species was a problem for those who believed that each individual was a manifestation of some immutable underlying ideal type or essence. Since natural theology presumes that a perfect creator created a perfect creation, these deviations from the ideal type were often regarded as consequences of the "sin" of Adam and Eve. Earlier transmutationists (such as Darwin's grandfather Erasmus) held that species change essence over time—for Darwin, species lacked any "essence" whatsoever! Because variation within populations is a necessary precondition for natural selection, Darwin goes to great length here to establish its existence. Darwin then shows here how breeders have used these variations to effect desired changes through a process of selective mating. Note that establishing existence of such phenomena is the first step in presenting a* vera causa *argument.]*

[. . .] It has been disputed at what period of time the causes of variability, whatever they may be, generally act; whether during the early or late period of development of the embryo, or at the instant of conception. Geoffroy St. Hilaire's experiments show that unnatural treatment of the embryo causes monstrosities; and monstrosities cannot be separated by any clear line of distinction from mere variations. But I am strongly inclined to suspect that the most frequent cause of variability may be attributed to the male and female reproductive elements having been affected prior to the act of conception.

[. . .] There are many laws regulating variation, some few of which can be dimly seen, and will be hereafter briefly mentioned. I will here only allude to what may be called correlation of growth. Any change in the embryo or larva will almost certainly entail changes in the mature animal. In monstrosities, the correlations between quite distinct parts are very curious; and many instances are given in Isidore Geoffroy St. Hilaire's great work on this subject. Breeders believe that long limbs are almost always accompanied by an elongated head. Some instances of correlation are quite whimsical; thus cats with blue eyes are invariably deaf; colour and constitutional peculiarities go together, of which many remarkable cases could be given amongst animals and plants. From the facts collected by Heusinger, it appears that white sheep and pigs are differently affected from coloured individuals by certain vegetable poisons. Hairless dogs have imperfect teeth; long-haired and coarse-haired animals are apt to have, as is asserted, long or many horns; pigeons with feathered feet have skin between their outer toes; pigeons with short beaks have small feet, and those with long beaks large feet. Hence, if man goes on selecting, and thus augmenting, any peculiarity, he will almost certainly unconsciously modify other parts of the structure, owing to the mysterious laws of the correlation of growth.

[. . .] Any variation which is not inherited is unimportant for us. But the number and diversity of inheritable deviations of structure, both those of slight and those of considerable physiological importance, is endless. [. . .] No breeder doubts how strong is the tendency to inheritance: like produces like is his fundamental belief: doubts have been thrown on this principle by theoretical writers alone. When a deviation appears not unfrequently, and we see it in the father and child, we cannot tell whether it may not be due to the same original cause acting on both; but when amongst individuals, apparently exposed to the same conditions, any very rare deviation, due to some extraordinary combination of circumstances, appears in the parent say, once amongst several million individuals and it reappears in the child, the mere doctrine of chances almost compels us to attribute its reappearance to inheritance. Every one must have heard of cases of albinism, prickly skin, hairy bodies, &c. appearing in several members of the same family. If strange and rare deviations of structure are truly inherited, less strange and commoner deviations may be freely admitted to be inheritable. Perhaps the correct way of viewing the whole subject, would be, to look at the inheritance of every character whatever as the rule, and non-inheritance as the anomaly. *[**Game note**: Why does Darwin want to characterize inheritance of traits as a general rule in his consideration of domestic animals? Primarily because by so doing, it will be easier to argue from analogy in the case of plants and animals in the state of nature in the subsequent chapter. **Character Alert**: Friend of Mill and the Disciple of Bacon especially will want to pay attention to Darwin's rhetorical strategy and all players should ponder the role of argument from analogy discussed in the Game book.]*

The laws governing inheritance are quite unknown;[8] no one can say why the same

---

8 Darwin had no clear notion of how inheritance worked at the level of physical processes. Darwin believed that "gemmules" of parents somehow blended in offspring; this fundamental misunderstanding severely weakened his theory and as P. Bowler [(1983) The eclipse of Darwinism. John Hopkins University Press: Baltimore, MD] has shown, natural selection had fallen out of favor by 1900 and was only resurrected following the rediscovery of Mendel's notion of particulate inheritance. Another interesting consequence of Darwin's misunderstanding was his lifelong belief in the Lamarckian notion of inheritance of acquired characters through habitual use or disuse.

Charles Darwin, the Copley Medal, and the Rise of Naturalism

peculiarity in different individuals of the same species, and in individuals of different species, is sometimes inherited and sometimes not so; why the child often reverts in certain characters to its grandfather or grandmother or other much more remote ancestor; why a peculiarity is often transmitted from one sex to both sexes or to one sex alone, more commonly but not exclusively to the like sex. A[n] . . . important rule . . . is that, at whatever period of life a peculiarity first appears, it tends to appear in the offspring at a corresponding age, though sometimes earlier. [. . . H]ereditary diseases and some other facts make me believe that the rule has a wider extension, and that when there is no apparent reason why a peculiarity should appear at any particular age, yet that it does tend to appear in the offspring at the same period at which it first appeared in the parent. I believe this rule to be of the highest importance in explaining the laws of embryology.

[. . .] there are hardly any domestic races, either amongst animals or plants, which have not been ranked by some competent judges as mere varieties, and by other competent judges as the descendants of aboriginally distinct species. If any marked distinction existed between domestic races and species, this source of doubt could not so perpetually recur.

On the Breeds of the Domestic Pigeon

*[**Game Note:** Pigeons were particularly important to the development of Darwin's ideas, so I have included rather extensive excerpts from the following section. Players in both A and X faction should read this section closely and may want to cite specific examples in Royal Society discussions.]*

Believing that it is always best to study some special group, I have, after deliberation, taken up domestic pigeons. I have kept every breed which I could purchase or obtain, and have been most kindly favoured with skins from several quarters of the world, more especially by the Hon. W. Elliot from India, and by the Hon. C. Murray from Persia. Many treatises in different languages have been published on pigeons, and some of them are very important, as being of considerably antiquity. I have associated with several eminent fanciers, and have been permitted to join two of the London Pigeon Clubs. The diversity of the breeds is something astonishing.[9] Compare the English carrier and the short-faced tumbler, and see the wonderful difference in their beaks, entailing corresponding differences in their skulls. The carrier, more especially the male bird, is also remarkable from the wonderful development of the carunculated skin about the head, and this is accompanied by greatly elongated eyelids, very large external orifices to the nostrils, and a wide gape of mouth. The short-faced tumbler has a beak in outline almost like that of a finch;

---

9 Many, if not most, of Darwin's mid-nineteenth century gentleman-readers would have been at least generally familiar with the astonishing diversity of various domestic pigeon breeds. Since most twenty-first century readers are not, I have included a Victorian-era illustration as a back cover for this abridgement.

and the common tumbler has the singular and strictly inherited habit of flying at a great height in a compact flock, and tumbling in the air head over heels. The runt is a bird of great size, with long, massive beak and large feet; some of the sub-breeds of runts have very long necks, others very long wings and tails, others singularly short tails. The barb is allied to the carrier, but, instead of a very long beak, has a very short and very broad one. The pouter has a much elongated body, wings, and legs; and its enormously developed crop, which it glories in inflating, may well excite astonishment and even laughter. The turbit has a very short and conical beak, with a line of reversed feathers down the breast; and it has the habit of continually expanding slightly the upper part of the oesophagus. The Jacobin has the feathers so much reversed along the back of the neck that they form a hood, and it has, proportionally to its size, much elongated wing and tail feathers. The trumpeter and laugher, as their names express, utter a very different coo from the other breeds. The fantail has thirty or even forty tail-feathers, instead of twelve or fourteen, the normal number in all members of the great pigeon family; and these feathers are kept expanded, and are carried so erect that in good birds the head and tail touch; the oil-gland is quite aborted. Several other less distinct breeds might have been specified.

In the skeletons of the several breeds, the development of the bones of the face in length and breadth and curvature differs enormously. The shape, as well as the breadth and length of the ramus of the lower jaw, varies in a highly remarkable manner. The number of the caudal and sacral vertebrae vary; as does the number of the ribs, together with their relative breadth and the presence of processes. The size and shape of the apertures in the sternum are highly variable; so is the degree of divergence and relative size of the two arms of the furcula. The proportional width of the gape of mouth, the proportional length of the eyelids, of the orifice of the nostrils, of the tongue (not always in strict correlation with the length of beak), the size of the crop and of the upper part of the oesophagus; the development and abortion of the oil-gland; the number of the primary wing and caudal feathers; the relative length of wing and tail to each other and to the body; the relative length of leg and of the feet; the number of scutellae on the toes, the development of skin between the toes, are all points of structure which are variable. The period at which the perfect plumage is acquired varies, as does the state of the down with which the nestling birds are clothed when hatched. The shape and size of the eggs vary. The manner of flight differs remarkably; as does in some breeds the voice and disposition. Lastly, in certain breeds, the males and females have come to differ to a slight degree from each other.

Altogether at least a score of pigeons might be chosen, which if shown to an ornithologist, and he were told that they were wild birds, would certainly, I think, be ranked by him as well-defined species. Moreover, I do not believe that any ornithologist would place the English carrier, the short-faced tumbler, the runt, the barb, pouter, and fantail in the same genus; more especially as in each of these breeds several truly-inherited sub-breeds, or species as he might have called them, could be shown him. [. . .] Great as the differences are between the breeds of pigeons, I am fully convinced that the common opinion of naturalists is correct, namely, that all have descended from the rock-pigeon (Columba livia), including under this term several geographical races or sub-species, which differ from each other in the most trifling respects. *[Game Note: Darwin then listed reasons which led him to this conclusion—if this is important for your character you may want to consult a full text version online.]*

[. . .] I have discussed the probable origin of domestic pigeons at some, yet quite insufficient, length; because when I first kept pigeons and watched the several kinds, knowing well how true they bred, I felt fully as much difficulty in believing that they could ever have descended from a common parent, as any naturalist could in coming to a similar conclusion in regard to the many species of finches, or other large groups of birds, in nature. One circumstance has struck me much; namely, that all the breeders of the various domestic animals and the cultivators of plants, with whom I have ever conversed, or whose treatises I have read, are firmly convinced that the several breeds to which each has attended, are descended from so many aboriginally distinct species. Ask, as I have asked, a celebrated raiser of Hereford cattle, whether his cattle might not have descended from long horns, and he will laugh you to scorn. I have never met a pigeon, or poultry, or duck, or rabbit fancier, who was not fully convinced that each main breed was descended from a distinct species. [. . .] Innumerable other examples could be given. The explanation, I think, is simple: from long-continued study they are strongly impressed with the differences between the several races; and though they well know that each race varies slightly, for they win their prizes by selecting such slight differences, yet they ignore all general arguments, and refuse to sum up in their minds slight differences accumulated during many successive generations. May not those naturalists who, knowing far less of the laws of inheritance than does the breeder, and knowing no more than he does of the intermediate links in the long lines of descent, yet admit that many of our domestic races have descended from the same parents may they not learn a lesson of caution, when they deride the idea of species in a state of nature being lineal descendants of other species? *[Game Note: Darwin's rhetoric here is intended to move fellow naturalists toward at least entertaining the possibility that natural selection might be sufficient to produce multiple species from a common ancestor. For some, however, his speculative tone in comments like this one marked a departure from the well-established inductive methods of science.]*

Selection

Let us now briefly consider the steps by which domestic races have been produced, either from one or from several allied species. Some little effect may, perhaps, be attributed to the direct action of the external conditions of life, and some little to habit; but he would be a bold man who would account by such agencies for the differences of a dray and race horse, a greyhound and bloodhound, a carrier and tumbler pigeon. One of the most remarkable features in our domesticated races is that we see in them adaptation, not indeed to the animal's or plant's own good, but to man's use or fancy. Some variations useful to him have probably arisen suddenly, or by one step; many botanists, for instance, believe that the fuller's teazle, with its hooks, which cannot be rivalled by any mechanical contrivance, is only a variety of the wild Dipsacus; and this amount of change may have suddenly arisen in a seedling. So it has probably been with the turnspit dog; and this is known to have been the case with the ancon sheep. But when we compare the dray-horse and race-horse, the dromedary and camel, the various breeds of sheep fitted either for cultivated land or mountain pasture, with the wool of one breed good for one purpose, and that of another breed for another purpose; when we compare the many breeds of dogs, each good for man in very different ways; when we compare the game-cock, so pertinacious in battle, with other breeds so little quarrelsome, with "everlasting layers" which never desire to sit, and with the bantam so small and elegant; when we compare the host of agricultural, culinary, orchard, and flower-garden races of plants, most useful to man at different seasons and for different purposes, or so beautiful in his eyes, we must, I think, look further than to mere variability. We cannot suppose that all the breeds were suddenly produced as perfect and as useful as we now see them; indeed, in several cases, we know that this has not been their history. The key is man's power of accumulative selection: nature gives successive variations; man adds them up in certain directions useful to him. In this sense he may be said to make for himself useful breeds.

The great power of this principle of selection is not hypothetical. It is certain that several of our eminent breeders have, even within a single lifetime, modified to a large extent some breeds of cattle and sheep. In order fully to realise what they have done, it is almost necessary to read several of the many treatises devoted to this subject, and to inspect the animals. Breeders habitually speak of an animal's organisation as something quite plastic, which they can model almost as they please. [. . .] That most skilful breeder, Sir John Sebright, used to say, with respect to pigeons, that "he would produce any given feather in three years, but it would take him six years to obtain head and beak." In Saxony the importance of the principle of selection in regard to merino sheep is so fully recognised, that men follow it as a trade: the sheep are placed on a table and are studied, like a picture by a connoisseur; this is done three times at intervals of months, and the sheep are each time marked and classed, so that the very best may ultimately be selected for breeding.

[. . .] It may be objected that the principle of selection has been reduced to methodical practice for scarcely more than three-quarters of a century; it has certainly been more attended to of late years, and many treatises have been published on the subject; and the result, has been, in a corresponding degree, rapid and important. But it is very far from true that the principle is a modern discovery. I could give several references to the full acknowledgement of the importance of the principle in works of high antiquity. In rude and barbarous periods of English history choice animals were often imported, and laws were passed to prevent their exportation: the destruction of horses under a certain size was ordered, and this may be compared to the "rouging" of plants by nurserymen. The principle of selection I find distinctly given in an ancient Chinese encyclopaedia. Explicit rules are laid down by some of the Roman classical writers. From passages in Genesis, it is clear that the colour of domestic animals was at that early period attended to. Savages now sometimes cross their dogs with wild canine animals, to improve the breed, and they formerly did so, as is attested by passages in Pliny. The savages in South Africa match their draught cattle by colour, as do some of the Esquimaux their teams of dogs. Livingstone shows how much good domestic breeds are valued by the negroes of the interior of Africa who have not associated with Europeans. Some of these facts do not show actual selection, but they show that the breeding of domestic animals was carefully attended to in ancient times, and is now attended to by the lowest savages. It would have been a strange fact, indeed, had attention not been paid to breeding, for the inheritance of good and bad qualities is so obvious.

At the present time, eminent breeders try by methodical selection, with a distinct object in view, to make a new strain or sub-breed, superior to anything existing in the country. But, for our purpose, a kind of Selection, which may be called Unconscious, and which results from every one trying to possess and breed from the best individual animals, is more important. Thus, a man who intends keeping pointers naturally tries to get as good dogs as he can, and afterwards breeds from his own best dogs, but he has no wish or expectation of permanently altering the breed.

[. . .] Youatt gives an excellent illustration of the effects of a course of selection, which may be considered as unconsciously followed, in so far that the breeders could never have expected or even have wished to have produced the result which ensued namely, the production of two distinct strains. The two flocks of Leicester sheep kept by Mr. Buckley and Mr. Burgess, as Mr. Youatt remarks, "have been purely bred from the original stock of Mr. Bakewell for upwards of fifty years. There is not a suspicion existing in the mind of any one at all acquainted with the subject that the owner of either of them has deviated in any one instance from the pure blood of Mr. Bakewell's flock, and yet the difference between the sheep possessed by these two gentlemen is so great that they have the appearance of being quite different varieties."

[. . .] Variability is governed by many unknown laws, more especially by that of correlation of growth. Something may be attributed to the direct action of the conditions of life. Something must be attributed to use and disuse. The final result is thus rendered infinitely complex. In some cases, I do not doubt that the intercrossing of species, aboriginally distinct, has played an important part in the origin of our domestic productions.[10] When in any country several domestic breeds have once been established, their occasional intercrossing, with the aid of selection, has, no doubt, largely aided in the formation of new sub-breeds; but the importance of the crossing of varieties has, I believe, been greatly exaggerated, both in regard to animals and to those plants which are propagated by seed. In plants which are temporarily propagated by cuttings, buds, &c., the importance of the crossing both of distinct species and of varieties is immense; for the cultivator here quite disregards the extreme variability both of hybrids and mongrels, and the frequent sterility of hybrids; but the cases of plants not propagated by seed are of little importance to us, for their endurance is only temporary. Over all these causes of change I am convinced that the accumulative action of selection, whether applied methodically and more quickly, or unconsciously and more slowly, but more efficiently, is by far the predominant power.

CHAPTER II

VARIATION UNDER NATURE

Variability – Individual differences – Doubtful species – Wide ranging, much diffused, and common species vary most – Species of the larger genera in any country vary more than the species of the smaller genera – Many of the species of the larger genera resemble varieties in being very closely, but unequally, related to each other, and in having restricted ranges

*[Game Note: In this chapter, Darwin establishes the unexpectedly high degree of variation present in populations in nature. He argues that differences among species are, in principle, similar to those commonly observed among sub-species and varieties. In many respects, **this is Darwin's key insight as it marks the transition from typological to population thinking** (see Mayr, E. 1976. Typological versus populational thinking. In: Mary, E. Evolution and the Diversity of Life. Harvard University Press: Cambridge UK.) Note also Darwin's comments on Lubbock's recent work.]*

Before applying the principles arrived at in the last chapter to organic beings in a state of nature, we must briefly discuss whether these latter are subject to any variation. To treat this subject at all properly, a long catalogue of dry facts should be

---

10 Recent work suggests that Darwin underestimated the importance of reconnecting or "anastomosing" lineages. In a particularly telling example, eukaryote cells contain mitochondrial endosymbionts descended from purple bacteria.

Charles Darwin, the Copley Medal, and the Rise of Naturalism

given; but these I shall reserve for my future work. Nor shall I here discuss the various definitions which have been given of the term species.[11] No one definition has as yet satisfied all naturalists; yet every naturalist knows vaguely what he means when he speaks of a species. Generally the term includes the unknown element of a distinct act of creation. The term "variety" is almost equally difficult to define; but here community of descent is almost universally implied, though it can rarely be proved. *[**Game Note:** Another implied but absent "long catalogue of dry facts." For A-Men, this can be taken as a further indication that Darwin's theory amounts to speculative hand-waving; for X-Men, absence of (abundant) evidence is not evidence of absence in principle but rather a call for further research. How you interpret such passages depends on your game objectives. Ultimately, the merit of Darwin's work and his viability as a candidate for the Copley Medal will hinge on the perceived balance of his inductive observations and deductive speculations].*

[. . .] Again, we have many slight differences which may be called individual differences, such as are known frequently to appear in the offspring from the same parents, or which may be presumed to have thus arisen, from being frequently observed in the individuals of the same species inhabiting the same confined locality. No one supposes that all the individuals of the same species are cast in the very same mould. These individual differences are highly important for us, as they afford materials for natural selection to accumulate, in the same manner as man can accumulate in any given direction individual differences in his domesticated productions. These individual differences generally affect what naturalists consider unimportant parts; but I could show by a long catalogue of facts, that parts which must be called important, whether viewed under a physiological or classificatory point of view, sometimes vary in the individuals of the same species. I am convinced that the most experienced naturalist would be surprised at the number of the cases of variability, even in important parts of structure, which he could collect on good authority, as I have collected, during a course of years. It should be remembered that systematists are far from pleased at finding variability in important characters, and that there are not many men who will laboriously examine internal and important organs, and compare them in many specimens of the same species. I should never have expected that the branching of the main nerves close to the great central ganglion of an insect would have been variable in the same species; I should have expected that changes of this nature could have been effected only by slow degrees: yet quite recently Mr. Lubbock has shown a degree of variability in these main nerves in Coccus, which may almost be compared to the irregular branching of the stem of a tree. This philosophical naturalist, has also quite recently shown that the muscles in the larvae of certain insects are very far from uniform. Authors sometimes argue in a circle when they state that important organs never vary; for

---

11 The "species problem" remains challenging; see e.g., M.C. McKitrick and R. M. Zink. 1988. Species concepts in ornithology. Condor 90: 1-13. No fewer than thirty definitions of the term species occur in the evolutionary literature and these are often contradictory. The once favored 'biological species concept' based on reproductive isolation has fallen out of favor in recent years as abundant counter-evidence has accumulated.

these same authors practically rank that character as important (as some few naturalists have honestly confessed) which does not vary; and, under this point of view, no instance of any important part varying will ever be found: but under any other point of view many instances assuredly can be given.[12]

[. . .] Certainly no clear line of demarcation has as yet been drawn between species and sub-species that is, the forms which in the opinion of some naturalists come very near to, but do not quite arrive at the rank of species; or, again, between sub-species and well-marked varieties, or between lesser varieties and individual differences. These differences blend into each other in an insensible series; and a series impresses the mind with the idea of an actual passage.

Hence I look at individual differences, though of small interest to the systematist, as of high importance for us, as being the first step towards such slight varieties as are barely thought worth recording in works on natural history. And I look at varieties which are in any degree more distinct and permanent, as steps leading to more strongly marked and more permanent varieties; and at these latter, as leading to sub-species, and to species. The passage from one stage of difference to another and higher stage may be, in some cases, due merely to the long-continued action of different physical conditions in two different regions; but I have not much faith in this view; and I attribute the passage of a variety, from a state in which it differs very slightly from its parent to one in which it differs more, to the action of natural selection in accumulating (as will hereafter be more fully explained) differences of structure in certain definite directions. Hence I believe a well-marked variety may be justly called an incipient species; but whether this belief be justifiable must be judged of by the general weight of the several facts and views given throughout this work.

Finally then varieties have the same general characters as species, for they cannot be distinguished from species, except, firstly, by the discovery of intermediate linking forms, and the occurrence of such links cannot affect the actual characters of the forms which they connect; and except, secondly, by a certain amount of difference, for two forms, if differing very little, are generally ranked as varieties, notwithstanding that intermediate linking forms have not been discovered; but the amount of difference considered necessary to give to two forms the rank of species is quite indefinite. In genera having more than the average number of species in any country, the species of these genera have more than the average number of varieties. In large genera the species are apt to be closely, but unequally, allied together, form-

---

12 Such arguments are called tautologies and natural selection itself, particularly in its misleading "survival of the fittest" formulation, has often and sometimes justifiably, been criticized as a tautology. For example, if we define fitness in terms of survival and reproduction and then attribute observed differences in survival and reproduction to differences in fitness, we have explained nothing. For a more detailed critique and solution to this problem see Mills, S.K. & Beatty, J.H. 1979. The propensity interpretation of fitness. Philosophy of Science 46: 263-268.

Charles Darwin, the Copley Medal, and the Rise of Naturalism

ing little clusters round certain species. Species very closely allied to other species apparently have restricted ranges. In all these several respects the species of large genera present a strong analogy with varieties. And we can clearly understand these analogies, if species have once existed as varieties, and have thus originated: whereas, these analogies are utterly inexplicable if each species has been independently created.

CHAPTER III

STRUGGLE FOR EXISTENCE

Bears on natural selection – The term used in a wide sense – Geometrical powers of increase – Rapid increase of naturalised animals and plants – Nature of the checks to increase – **Competition universal** – Effects of climate – Protection from the number of individuals – Complex relations of all animals and plants throughout nature – Struggle for life most severe between individuals and varieties of the same species; often severe between species of the same genus – The relation of organism to organism the most important of all relations

[**Game Note:** *Chapter 3 is critical for an understanding of Darwin's theory and it would be well worth your time to peruse the full text of this chapter on-line. Pay attention to the way in which struggle for existence is related to geometrical increases in population and thus to seemingly inevitable competition. Although Darwin notes that he uses the term struggle for existence in "a large and metaphorical sense" and indeed provides numerous examples where struggle is related not to competition but to predation, parasitism, or environmental conditions, he nonetheless adopts the then prevailing Malthusian view that competition is universal. In effect, he reads an interpretation of human culture back into nature in spite of his own evidence to the contrary. As a result, many readers presume (wrongly) that natural selection requires competition among individuals. More recent work suggests that "surplus reproduction" is an evolutionary response to high mortality rates due to pathogens, parasites, predators, and environmental events and that competition a relatively rare, negative-negative ecological interaction that tends to be eliminated by natural selection.[13]*

Character Alert: *All players—but especially the Malthusian Mathematician and the Civil Engineer—should pay particular attention to Darwin's emphasis on a "universal struggle for life"—for Darwin and his contemporaries human events*

---

13 For more on this see Connell, J.H. 1980. Diversity and the co-evolution of competitors, or the ghost of competition past. Oikos 35: 131-138; also Hairston, N.G., Smith, F.E., and Slobodkin, L.B. 1960. Community structure, population control, and competition. American Naturalist 94: 421-425.

such as the Irish potato famines of the late 1840s demonstrated that even if "strug-
gles for existence" were intermittent, they were nonetheless profoundly important
for understanding differences in survival among populations. For example, many
members of the Royal Society believed that the potato famines killed so many Irish
because of the innate inferiority of the Celtic race relative to the Saxons (e.g., Knox,
R. 1850).]

Before entering on the subject of this chapter, I must make a few preliminary remarks, to show how the struggle for existence bears on Natural Selection. It has been seen in the last chapter that amongst organic beings in a state of nature there is some individual variability; indeed I am not aware that this has ever been disputed. It is immaterial for us whether a multitude of doubtful forms be called species or sub-species or varieties; what rank, for instance, the two or three hundred doubtful forms of British plants are entitled to hold, if the existence of any well-marked varieties be admitted. But the mere existence of individual variability and of some few well-marked varieties, though necessary as the foundation for the work, helps us but little in understanding how species arise in nature. How have all those exquisite adaptations of one part of the organisation to another part, and to the conditions of life, and of one distinct organic being to another being, been perfected? We see these beautiful co-adaptations most plainly in the woodpecker and missletoe; and only a little less plainly in the humblest parasite which clings to the hairs of a quadruped or feathers of a bird; in the structure of the beetle which dives through the water; in the plumed seed which is wafted by the gentlest breeze; in short, we see beautiful adaptations everywhere and in every part of the organic world.

[. . .] Nothing is easier than to admit in words the truth of the universal struggle for life, or more difficult at least I have found it so than constantly to bear this conclusion in mind. Yet unless it be thoroughly engrained in the mind, I am convinced that the whole economy of nature, with every fact on distribution, rarity, abundance, extinction, and variation, will be dimly seen or quite misunderstood. We behold the face of nature bright with gladness, we often see superabundance of food; we do not see, or we forget, that the birds which are idly singing round us mostly live on insects or seeds, and are thus constantly destroying life; or we forget how largely these songsters, or their eggs, or their nestlings are destroyed by birds and beasts of prey; we do not always bear in mind, that though food may be now superabundant, it is not so at all seasons of each recurring year.

I should premise that I use the term Struggle for Existence in a large and metaphorical sense, including dependence of one being on another, and including (which is more important) not only the life of the individual, but success in leaving progeny. Two canine animals in a time of dearth, may be truly said to struggle with each other which shall get food and live. But a plant on the edge of a desert is said to struggle for life against the drought, though more properly it should be said to be dependent on the moisture. A plant which annually produces a thousand seeds, of which on an average only one comes to maturity, may be more truly said to struggle with the plants of the same and other kinds which already clothe the ground. The missletoe is dependent on the apple and a few other trees, but can only in a far-fetched sense be said to struggle with these trees, for if too many of these parasites grow on the same tree, it will languish and die. But several seedling missletoes, growing close together on the same branch, may more truly be said to struggle with each other. As the missletoe is disseminated by birds, its existence depends on birds; and it may metaphorically be said to struggle with other fruit-bearing plants, in order to tempt birds to devour and thus disseminate its seeds rather than those of other plants. In these several senses, which pass into each other, I use for convenience sake the general term of struggle for existence. *[Game Note: Some critical analysis of Darwin's "Struggle for Existence" metaphor might pay dividends in the context of the game—for either faction or for careful non-partisans seeking to advance their case on ancillary issues (i.e., race, gender, social welfare).]*

A struggle for existence inevitably follows from the high rate at which all organic beings tend to increase.[14] Every being, which during its natural lifetime produces several eggs or seeds, must suffer destruction during some period of its life, and during some season or occasional year, otherwise, on the principle of geometrical increase, its numbers would quickly become so inordinately great that no country could support the product. Hence, as more individuals are produced than can possibly survive, there must in every case be a struggle for existence, either one individual with another of the same species, or with the individuals of distinct species, or with the physical conditions of life. It is the doctrine of Malthus applied with manifold force to the whole animal and vegetable kingdoms; for in this case there can be no artificial increase of food, and no prudential restraint from marriage. Although some species may be now increasing, more or less rapidly, in numbers, all cannot do so, for the world would not hold them.[15] *[Character Alert: Malthusian Mathematician and Civil Engineer.]*

---

14 It can be argued (and this is an emerging view among 21st-century evolutionary ecologists) that observed high reproductive rates are an effect rather than a cause of struggle for existence in the face of predators, parasites, pathogens and environmental caprice. Note that destruction of eggs or seeds in no way implies competition as a necessary fate. The "doctrine of Malthus" was derived not from "nature" but from interpretations of human experience in industrial and imperial nineteenth century English culture. D.H. Abbott [(1988) Natural suppression of fertility. Zoological Journal of the Linnean Society 60: 7-28] documents numerous cases where individual reproduction is not maximized in natural systems. Others have shown that reproductive tactics are quite plastic and can be adjusted in response to various environmental cues (see e.g., D.P. Siems and R.S. Sikes (1995) Tactical trade-offs between growth and reproduction. Environmental Biology of Fishes 53: 319-329).

15 Reverend Thomas Malthus wrote "An essay on the principle of population as it affects the future improvement of society" in 1798 (Johnson: London). This work was an attack on "the Poor Laws," early attempts to provide a economic safety net for industrial England's economically disenfranchised. For Malthus, poverty was inevitable and was part of God's inscrutable plan—providing for the welfare of impoverished classes would only aggravate problems by encouraging higher birth rates among inferior stock. Darwin, who read for amusement a sixth edition (1828) copy of Malthus in 1838, was a generous contributor to various progressive social causes and thus found the views of Malthus morally repugnant. Nonetheless, he rather uncritically adopted the (mistaken) Malthusian notion that food supplies increase arithmetically while populations increase geometrically.

There is no exception to the rule that every organic being naturally increases at so high a rate, that if not destroyed, the earth would soon be covered by the progeny of a single pair. Even slow-breeding man has doubled in twenty-five years, and at this rate, in a few thousand years, there would literally not be standing room for his progeny.[16] Linnaeus has calculated that if an annual plant produced only two seeds and there is no plant so unproductive as this and their seedlings next year produced two, and so on, then in twenty years there would be a million plants. The elephant is reckoned to be the slowest breeder of all known animals, and I have taken some pains to estimate its probable minimum rate of natural increase: it will be under the mark to assume that it breeds when thirty years old, and goes on breeding till ninety years old, bringing forth three pairs of young in this interval; if this be so, at the end of the fifth century there would be alive fifteen million elephants, descended from the first pair *[**Game Note:** Elephant example provides a useful paper and pencil demonstration lab exercise especially for the X-man faction—consult with your Gamemaster on how to conduct such a demonstration].*

[. . .] In looking at Nature, it is most necessary to keep the foregoing considerations always in mind never to forget that every single organic being around us may be said to be striving to the utmost to increase in numbers; that each lives by a struggle at some period of its life; that heavy destruction inevitably falls either on the young or old, during each generation or at recurrent intervals. Lighten any check, mitigate the destruction ever so little, and the number of the species will almost instantaneously increase to any amount. The face of Nature may be compared to a yielding surface, with ten thousand sharp wedges packed close together and driven inwards by incessant blows, sometimes one wedge being struck, and then another with greater force.

What checks the natural tendency of each species to increase in number is most obscure. Look at the most vigorous species; by as much as it swarms in numbers, by so much will its tendency to increase be still further increased. We know not exactly what the checks are in even one single instance. [. . .] Here I will make only a few remarks, just to recall to the reader's mind some of the chief points. Eggs or very young animals seem generally to suffer most, but this is not invariably the case. With plants there is a vast destruction of seeds, [. . .] Seedlings, also, are destroyed in vast numbers by various enemies; for instance, on a piece of ground three feet long and two wide, dug and cleared, and where there could be no choking from other plants, I marked all the seedlings of our native weeds as they came up, and out of the 357 no less than 295 were destroyed, chiefly by slugs and insects. *[**Game Note:** Here is an excellent empirical example you may wish to cite in your*

---

16 The human population of industrializing England had doubled in a quarter century with much of the increase occurring within the poorest classes of newly urbanized workers. In 1825 London was already the world's largest city with a population of 1.35 million; by 1875 London's population had reached 4.25 million. For a sense of living conditions in London, see http://www.fidnet.com/~dap1955/dickens/dickens_london.html.

debates; it is worth noting however that observed 83% mortality rate was due pri-marily to herbivory rather than competition for resources.]

[. . .] The amount of food for each species of course gives the extreme limit to which each can increase; but very frequently it is not the obtaining food, but the serving as prey to other animals, which determines the average numbers of a species.

[. . .] Climate plays an important part in determining the average numbers of a species, and periodical seasons of extreme cold or drought, I believe to be the most effective of all checks. I estimated that the winter of 1854-55 destroyed four-fifths of the birds in my own grounds; and this is a tremendous destruction, when we remember that ten per cent. is an extraordinarily severe mortality from epidemics with man. The action of climate seems at first sight to be quite independent of the struggle for existence; but in so far as climate chiefly acts in reducing food, it brings on the most severe struggle between the individuals, whether of the same or of distinct species, which subsist on the same kind of food. Even when climate, for instance extreme cold, acts directly, it will be the least vigorous, or those which have got least food through the advancing winter, which will suffer most. [. . .] [17]

[. . .] When a species . . . increases inordinately in numbers in a small tract, epi-demics—at least this seems generally to occur with our game animals—often ensue: and here we have a limiting check independent of the struggle for life. But even some of these so-called epidemics appear to be due to parasitic worms, which have from some cause, possibly in part through facility of diffusion amongst the crowded animals, been disproportionably favoured: and here comes in a sort of struggle between the parasite and its prey.

[. . .] Many cases are on record showing how complex and unexpected are the checks and relations between organic beings, which have to struggle together in the same country. I will give only a single instance, which, though a simple one, has interested me. In Staffordshire, on the estate of a relation where I had ample means of investigation, there was a large and extremely barren heath, which had never been touched by the hand of man; but several hundred acres of exactly the same nature had been enclosed twenty-five years previously and planted with Scotch fir. The change in the native vegetation of the planted part of the heath was most remarkable, more than is generally seen in passing from one quite different soil to

---

17 Winter "starvation" in white-tail deer provides an interesting example. Artificially high deer population lev-els (intentionally achieved as a management goal) coupled with reductions of natural predators creates a situ-ation where high mortality occurs during cold, snowy winters. Although overall food supplies remain adequate, snow depth greatly reduces mobility resulting in local depletion of food and increased death rates among youngest and oldest sectors of the population. Most deaths are a result of exposure, secondary infections, and predation.

another: not only the proportional numbers of the heath-plants were wholly changed, but twelve species of plants (not counting grasses and carices) flourished in the plantations, which could not be found on the heath. The effect on the insects must have been still greater, for six insectivorous birds were very common in the plantations, which were not to be seen on the heath; and the heath was frequented by two or three distinct insectivorous birds. Here we see how potent has been the effect of the introduction of a single tree, nothing whatever else having been done, with the exception that the land had been enclosed, so that cattle could not enter. But how important an element enclosure is, I plainly saw near Farnham, in Surrey. Here there are extensive heaths, with a few clumps of old Scotch firs on the distant hill-tops: within the last ten years large spaces have been enclosed, and self-sown firs are now springing up in multitudes, so close together that all cannot live. When I ascertained that these young trees had not been sown or planted, I was so much surprised at their numbers that I went to several points of view, whence I could examine hundreds of acres of the unenclosed heath, and literally I could not see a single Scotch fir, except the old planted clumps. But on looking closely between the stems of the heath, I found a multitude of seedlings and little trees, which had been perpetually browsed down by the cattle. In one square yard, at a point some hundreds yards distant from one of the old clumps, I counted thirty-two little trees; and one of them, judging from the rings of growth, had during twenty-six years tried to raise its head above the stems of the heath, and had failed. No wonder that, as soon as the land was enclosed, it became thickly clothed with vigorously grow-ing young firs. Yet the heath was so extremely barren and so extensive that no one would ever have imagined that cattle would have so closely and effectually searched it for food.

Not that in nature the relations can ever be as simple as this. Battle within battle must ever be recurring with varying success; and yet in the long-run the forces are so nicely balanced, that the face of nature remains uniform for long periods of time, though assuredly the merest trifle would often give the victory to one organic being over another. Nevertheless so profound is our ignorance, and so high our presump-tion, that we marvel when we hear of the extinction of an organic being; and as we do not see the cause, we invoke cataclysms to desolate the world, or invent laws on the duration of the forms of life!

[. . .] Look at a plant in the midst of its range, why does it not double or quadru-ple its numbers? We know that it can perfectly well withstand a little more heat or cold, dampness or dryness, for elsewhere it ranges into slightly hotter or colder, damper or drier districts. In this case we can clearly see that if we wished in imag-ination to give the plant the power of increasing in number, we should have to give it some advantage over its competitors, or over the animals which preyed on it. On the confines of its geographical range, a change of constitution with respect to cli-mate would clearly be an advantage to our plant; but we have reason to believe that only a few plants or animals range so far, that they are destroyed by the rigour of the climate alone. Not until we reach the extreme confines of life, in the arctic regions or on the borders of an utter desert, will competition cease.[18]

It is good thus to try in our imagination to give any form some advantage over another. Probably in no single instance should we know what to do, so as to succeed. It will convince us of our ignorance on the mutual relations of all organic beings; a conviction as necessary, as it seems to be difficult to acquire. All that we can do, is to keep steadily in mind that each organic being is striving to increase at a geometrical ratio; that each at some period of its life, during some season of the year, during each generation or at intervals, has to struggle for life, and to suffer great destruction. When we reflect on this struggle, we may console ourselves with the full belief that the war of nature is not incessant, that no fear is felt, that death is generally prompt, and that the vigorous, the healthy, and the happy survive and multiply.

CHAPTER IV

NATURAL SELECTION

Natural Selection – its power compared with man's selection – its power on characters of trifling importance – its power at all ages and on both sexes – **Sexual Selection** – On the generality of intercrosses between individuals of the same species – Circumstances favourable and unfavourable to Natural Selection, namely, intercrossing, isolation, number of individuals – Slow action – Extinction caused by Natural Selection – **Divergence of Character,** related to the diversity of inhabitants of any small area, and to naturalisation – Action of Natural Selection, through Divergence of Character and Extinction, on the descendants from a common parent – **Explains the Grouping of all organic beings**

[Game Note: This chapter presents the core idea of Darwin's theory of natural selection. The theory is deductively certain provided that its premises are correct so you should try to pick out both the logical structure of the argument and the nature of Darwin's assumptions. Read critically and look for places where assumptions seem dubious. Character Alert: understanding this chapter is critical for all players!]

How will the struggle for existence, discussed too briefly in the last chapter, act in regard to variation? Can the principle of selection, which we have seen is so potent

---

18 Darwin's examples in this chapter primarily focus on effects of predation and disease but here he makes a (culturally-induced?) shift in focus and concludes that competition ceases only at the "extreme confines of life." Among his cited examples, competition is only apparent among firs within human constructed exclosures!

in the hands of man, apply in nature? I think we shall see that it can act most effectually. Let it be borne in mind in what an endless number of strange peculiarities our domestic productions, and, in a lesser degree, those under nature, vary; and how strong the hereditary tendency is. Under domestication, it may be truly said that the, whole organisation becomes in some degree plastic. Let it be borne in mind how infinitely complex and close-fitting are the mutual relations of all organic beings to each other and to their physical conditions of life. Can it, then, be thought improbable, seeing that variations useful to man have undoubtedly occurred, that other variations useful in some way to each being in the great and complex battle of life, should sometimes occur in the course of thousands of generations? If such do occur, can we doubt (remembering that many more individuals are born than can possibly survive) that individuals having any advantage, however slight, over others, would have the best chance of surviving and of procreating their kind? On the other hand, we may feel sure that any variation in the least degree injurious would be rigidly destroyed. This preservation of favourable variations and the rejection of injurious variations, I call Natural Selection. Variations neither useful nor injurious would not be affected by natural selection, and would be left a fluctuating element, as perhaps we see in the species called polymorphic.[19]

We shall best understand the probable course of natural selection by taking the case of a country undergoing some physical change, for instance, of climate. The proportional numbers of its inhabitants would almost immediately undergo a change, and some species might become extinct. We may conclude, from what we have seen of the intimate and complex manner in which the inhabitants of each country are bound together, that any change in the numerical proportions of some of the inhabitants, independently of the change of climate itself, would most seriously affect many of the others. If the country were open on its borders, new forms would certainly immigrate, and this also would seriously disturb the relations of some of the former inhabitants. Let it be remembered how powerful the influence of a single introduced tree or mammal has been shown to be. But in the case of an island, or of a country partly surrounded by barriers, into which new and better adapted forms could not freely enter, we should then have places in the economy of nature which would assuredly be better filled up, if some of the original inhabitants were in some manner modified; for, had the area been open to immigration, these same places would have been seized on by intruders. In such case, every slight modification, which in the course of ages chanced to arise, and which in any way favoured the individuals of any of the species, by better adapting them to their altered conditions, would tend to be preserved; and natural selection would thus have free scope for the work of improvement.

---

19 Recent work suggests that much of the genetic and biochemical variation in populations is in fact neutral with respect to natural selection. For more on this see Kimura, M. 1989. The neutral theory of molecular evolution and the worldview of the neutralists. Genome 31: 24-31. Random changes in the frequencies of different forms of a given gene (i.e., alleles) are referred to as genetic drift. Genetic drift inevitably produces changes in form and physiology that are not related to any particular adaptive advantages.

[. . .] It may be said that natural selection is daily and hourly scrutinising, through-out the world, every variation, even the slightest; rejecting that which is bad, pre-serving and adding up all that is good; silently and insensibly working, whenever and wherever opportunity offers, at the improvement of each organic being in rela-tion to its organic and inorganic conditions of life. We see nothing of these slow changes in progress, until the hand of time has marked the long lapses of ages, and then so imperfect is our view into long past geological ages, that we only see that the forms of life are now different from what they formerly were.

Sexual Selection

[Character Alert: Paleontologist and Astronomer] . . . this leads me to say a few words on what I call Sexual Selection. This depends, not on a struggle for exis-tence, but on a struggle between the males for possession of the females; the result is not death to the unsuccessful competitor, but few or no offspring. Sexual selec-tion is, therefore, less rigorous than natural selection. Generally, the most vigorous males, those which are best fitted for their places in nature, will leave most proge-ny. But in many cases, victory will depend not on general vigour, but on having special weapons, confined to the male sex. [. . .] Amongst birds, the contest is often of a more peaceful character. All those who have attended to the subject, believe that there is the severest rivalry between the males of many species to attract by singing the females. The rock-thrush of Guiana, birds of paradise, and some oth-ers, congregate; and successive males display their gorgeous plumage and perform strange antics before the females, which standing by as spectators, at last choose the most attractive partner. [. . .] It may appear childish to attribute any effect to such apparently weak means: I cannot here enter on the details necessary to sup-port this view; but if man can in a short time give elegant carriage and beauty to his bantams, according to his standard of beauty, I can see no good reason to doubt that female birds, by selecting, during thousands of generations, the most melodi-ous or beautiful males, according to their standard of beauty, might produce a marked effect.[20]

Illustrations of the action of Natural Selection

In order to make it clear how, as I believe, natural selection acts, I must beg permis-sion to give one or two imaginary illustrations. Let us take the case of a wolf, which preys on various animals, securing some by craft, some by strength, and some by fleetness; and let us suppose that the fleetest prey, a deer for instance, had

---

20 Because Darwin's notion of sexual selection included active female choice as an agent every bit as impor-tant as male-male contest competition, it was a difficult idea for even his supporters to accept. Recent DNA-based paternity studies in several species reveal that winners of male-male competition do not necessarily end up fathering more offspring and also suffer reduced life-spans due directly to injury and indirectly to stress related physiological problems [see e.g., R. M. Sapolsky. (1990) Stress in the wild. Scientific American. Jan: 116-123].

from any change in the country increased in numbers, or that other prey had decreased in numbers, during that season of the year when the wolf is hardest pressed for food. I can under such circumstances see no reason to doubt that the swiftest and slimmest wolves would have the best chance of surviving, and so be preserved or selected, provided always that they retained strength to master their prey at this or at some other period of the year, when they might be compelled to prey on other animals. I can see no more reason to doubt this, than that man can improve the fleetness of his greyhounds by careful and methodical selection, or by that unconscious selection which results from each man trying to keep the best dogs without any thought of modifying the breed.

*[**Game Note:** Darwin next presents several hypothetical cases of how natural selection might be expected to operate. Because some of these examples were overly simplistic they were subject to great skepticism and occasional ridicule. **Character Alert:** A-Men should consult the full text to glean some of these examples as illustrations of Darwin's speculative deviations from the time-proven, inductive methods of science.]*

I am well aware that this doctrine of natural selection, exemplified in the above imaginary instances, is open to the same objections which were at first urged against Sir Charles Lyell's noble views on "the modern changes of the earth, as illustrative of geology"; but we now very seldom hear the action, for instance, of the coast-waves, called a trifling and insignificant cause, when applied to the excavation of gigantic valleys or to the formation of the longest lines of inland cliffs. Natural selection can act only by the preservation and accumulation of infinitesimally small inherited modifications, each profitable to the preserved being; and as modern geology has almost banished such views as the excavation of a great valley by a single diluvial wave, so will natural selection, if it be a true principle, banish the belief of the continued creation of new organic beings, or of any great and sudden modification in their structure.

*[**Game Note:** Next follows an extended digression on inter-crossing that was not viewed central to Darwin's main argument. Following that Darwin offers a rather rambling section describing circumstances supposed to be favorable for natural selection; only part of his summary is included here.]*

That natural selection will always act with extreme slowness, I fully admit. Its action depends on there being places in the polity of nature, which can be better occupied by some of the inhabitants of the country undergoing modification of some kind. The existence of such places will often depend on physical changes, which are generally very slow, and on the immigration of better adapted forms having been checked. But the action of natural selection will probably still oftener depend on some of the inhabitants becoming slowly modified; the mutual relations of many of the other inhabitants being thus disturbed. Nothing can be effected, unless favourable variations occur, and variation itself is apparently always a very slow process. [. . .] Many will exclaim that these several causes are amply sufficient wholly to stop the action of natural selection. I do not believe so. On the other hand, I do believe that natural selection will always act very slowly, often only at long intervals of time, and generally on only a very few of the inhabitants of the same region at the same time. I further believe, that this very slow, intermittent action of natural selection accords perfectly well with what geology tells us of the rate and manner at which the inhabitants of this world have changed.

Slow though the process of selection may be, if feeble man can do much by his powers of artificial selection, I can see no limit to the amount of change, to the beauty and infinite complexity of the coadaptations between all organic beings, one with another and with their physical conditions of life, which may be effected in the long course of time by nature's power of selection.[21]

Extinction caused by Natural Selection

[. . .] Natural selection acts solely through the preservation of variations in some way advantageous, which consequently endure. But as from the high geometrical powers of increase of all organic beings, each area is already fully stocked with inhabitants, it follows that as each selected and favoured form increases in number, so will the less favoured forms decrease and become rare. Rarity, as geology tells us, is the precursor to extinction. We can, also, see that any form represented by few individuals will, during fluctuations in the seasons or in the number of its enemies, run a good chance of utter extinction. But we may go further than this; for as new forms are continually and slowly being produced, unless we believe that the number of specific forms goes on perpetually and almost indefinitely increasing, numbers inevitably must become extinct.[22]

Divergence of Character

The principle, which I have designated by this term, is of high importance on my theory, and explains, as I believe, several important facts. In the first place, varieties, even strongly-marked ones, though having somewhat of the character of species as is shown by the hopeless doubts in many cases how to rank them yet cer-

---

21 Darwin's debt to Lyell and other geologist is apparent here. For an extended analysis of the importance of the emergence of the notion of "deep time" see Gould, S.J. 1989. Time's arrow, time's cycle. Harvard University Press: Boston MA.

22 The idea that competitive interactions lead to extinctions is known as competitive exclusion. The formal mathematical theory of competitive exclusion was developed by G.F. Gause (1934) in a "The Struggle for Existence," (Williams and Wilkins: Baltimore, MD). After numerous failed attempts, Gause was eventually able to develop a laboratory system with micro-organisms that produced results consistent with his theory. However, competitive exclusion has never been conclusively demonstrated in any natural system!

tainly differ from each other far less than do good and distinct species.[23] Nevertheless, according to my view, varieties are species in the process of formation, or are, as I have called them, incipient species. How, then, does the lesser difference between varieties become augmented into the greater difference between species? That this does habitually happen, we must infer from most of the innumerable species throughout nature presenting well-marked differences; whereas varieties, the supposed prototypes and parents of future well-marked species, present slight and ill-defined differences. Mere chance, as we may call it, might cause one variety to differ in some character from its parents, and the offspring of this variety again to differ from its parent in the very same character and in a greater degree; but this alone would never account for so habitual and large an amount of difference as that between varieties of the same species and species of the same genus.

[. . .] The truth of the principle, that the greatest amount of life can be supported by great diversification of structure, is seen under many natural circumstances. In an extremely small area, especially if freely open to immigration, and where the contest between individual and individual must be severe, we always find great diversity in its inhabitants. For instance, I found that a piece of turf, three feet by four in size, which had been exposed for many years to exactly the same conditions, supported twenty species of plants, and these belonged to eighteen genera and to eight orders, which shows how much these plants differed from each other. So it is with the plants and insects on small and uniform islets; and so in small ponds of fresh water. Farmers find that they can raise most food by a rotation of plants belonging to the most different orders: nature follows what may be called a simultaneous rotation. Most of the animals and plants which live close round any small piece of ground, could live on it (supposing it not to be in any way peculiar in its nature), and may be said to be striving to the utmost to live there; but, it is seen, that where they come into the closest competition with each other, the advantages of diversification of structure, with the accompanying differences of habit and constitution, determine that the inhabitants, which thus jostle each other most closely, shall, as a general rule, belong to what we call different genera and orders.[24]

The accompanying diagram[25] will aid us in understanding this rather perplexing subject. Let A to L represent the species of a genus large in its own country; these

---

23 Character displacement (W.L. Brown and E.O. Wilson. 1956. Character Displacement. Systematic Zoology 5: 49-64) is the usual result of a range overlap between species with similar resource requirements. Here species diverge, over a few generations, in ways that minimize competition for shared resources. For a more recent analysis see P.A. Abrams (1986) Character displacement and niche shift analyzed using consumer resource models of competition. Theoretical Population Biology 29: 107-159.

24 This is a fascinating conjecture, to my knowledge so far untested. Potentially great dissertation topic!

25 The diagram mentioned here is the only illustration in the "Origin" and it is reproduced inside the front cover of the present abridgement. Darwin presented an extended discussion of this diagram that is worth careful study. In the interest of brevity I have omitted most of that text here but curious readers are encouraged to consult one of the on-line versions of noted earlier.

Charles Darwin, the Copley Medal, and the Rise of Naturalism

species are supposed to resemble each other in unequal degrees, as is so generally the case in nature, and as is represented in the diagram by the letters standing at unequal distances. I have said a large genus, because we have seen in the second chapter, that on an average more of the species of large genera vary than of small genera; and the varying species of the large genera present a greater number of varieties. We have, also, seen that the species, which are the commonest and the most widely-diffused, vary more than rare species with restricted ranges. Let (A) be a common, widely-diffused, and varying species, belonging to a genus large in its own country. The little fan of diverging dotted lines of unequal lengths proceeding from (A), may represent its varying offspring. The variations are supposed to be extremely slight, but of the most diversified nature; they are not supposed all to appear simultaneously, but often after long intervals of time; nor are they all supposed to endure for equal periods. Only those variations which are in some way profitable will be preserved or naturally selected. And here the importance of the principle of benefit being derived from divergence of character comes in; for this will generally lead to the most different or divergent variations (represented by the outer dotted lines) being preserved and accumulated by natural selection. When a dotted line reaches one of the horizontal lines, and is there marked by a small numbered letter, a sufficient amount of variation is supposed to have been accumulated to have formed a fairly well-marked variety, such as would be thought worthy of record in a systematic work. [. . .]

Summary of Chapter

If during the long course of ages and under varying conditions of life, organic beings vary at all in the several parts of their organisation, and I think this cannot be disputed; if there be, owing to the high geometrical powers of increase of each species, at some age, season, or year, a severe struggle for life, and this certainly cannot be disputed; then, considering the infinite complexity of the relations of all organic beings to each other and to their conditions of existence, causing an infinite diversity in structure, constitution, and habits, to be advantageous to them, I think it would be a most extraordinary fact if no variation ever had occurred useful to each being's own welfare, in the same way as so many variations have occurred useful to man. But if variations useful to any organic being do occur, assuredly individuals thus characterised will have the best chance of being preserved in the struggle for life; and from the strong principle of inheritance they will tend to produce offspring similarly characterised. This principle of preservation, I have called, for the sake of brevity, Natural Selection. Natural selection, on the principle of qualities being inherited at corresponding ages, can modify the egg, seed, or young, as easily as the adult. Amongst many animals, sexual selection will give its aid to ordinary selection, by assuring to the most vigorous and best adapted males the greatest number of offspring. Sexual selection will also give characters useful to the males alone, in their struggles with other males.

The affinities of all the beings of the same class have sometimes been represented by a great tree.[26] I believe this simile largely speaks the truth. The green and budding twigs may represent existing species; and those produced during each former year may represent the long succession of extinct species. At each period of growth all the growing twigs have tried to branch out on all sides, and to overtop and kill the surrounding twigs and branches, in the same manner as species and groups of species have tried to overmaster other species in the great battle for life. The limbs divided into great branches, and these into lesser and lesser branches, were themselves once, when the tree was small, budding twigs; and this connexion of the former and present buds by ramifying branches may well represent the classification of all extinct and living species in groups subordinate to groups. Of the many twigs which flourished when the tree was a mere bush, only two or three, now grown into great branches, yet survive and bear all the other branches; so with the species which lived during long-past geological periods, very few now have living and modified descendants. From the first growth of the tree, many a limb and branch has decayed and dropped off; and these lost branches of various sizes may represent those whole orders, families, and genera which have now no living representatives, and which are known to us only from having been found in a fossil state. [. . .] As buds give rise by growth to fresh buds, and these, if vigorous, branch out and overtop on all sides many a feebler branch, so by generation I believe it has been with the great Tree of Life, which fills with its dead and broken branches the crust of the earth, and covers the surface with its ever branching and beautiful ramifications.

CHAPTER V

LAWS OF VARIATION

Effects of external conditions – Use and disuse, combined with natural selection; organs of flight and of vision – Acclimatisation – Correlation of growth – Compensation and economy of growth – False correlations – Multiple, rudimentary, and lowly organised structures variable – Parts developed in an unusual manner are highly variable: specific character more variable than generic: secondary sexual characters variable – Species of the same genus vary in an analogous manner – Reversions to long-lost characters – **Summary**

---

26 The great tree of life has been called the 'dominant visual metaphor' in evolutionary biology. Inside the back cover, I have provided a representative 1874 image of a "Lebensbaum" from the German biologist Ernst Haeckel (1834-1919), the self-described "apostle of Darwin in Germany." Note that the image can be as misleading as it can be informative. The position of man at the pinnacle of the tree has no biological justification. On a related note, the "Tree of Life" project (http://tolweb.org/tree/phylogeny.html), a distributed attempt to provide a systematic overview of earth's diverse biota, is one of the more interesting and useful internet applications.

[Game Note: Darwin's misunderstanding of genetics posed perhaps the greatest single weakness of his theory. Consequently, both his adversaries and his supporters will want to pay close attention to this chapter.] Our ignorance of the laws of variation is profound. [. . .] The external conditions of life, as climate and food, &c., seem to have induced some slight modifications. Habit in producing constitutional differences, and use in strengthening, and disuse in weakening and diminishing organs, seem to have been more potent in their effects. Homologous parts tend to vary in the same way, and homologous parts tend to cohere. Modifications in hard parts and in external parts sometimes affect softer and internal parts. When one part is largely developed, perhaps it tends to draw nourishment from the adjoining parts; and every part of the structure which can be saved without detriment to the individual, will be saved. Changes of structure at an early age will generally affect parts subsequently developed; and there are very many other correlations of growth, the nature of which we are utterly unable to understand. Multiple parts are variable in number and in structure, perhaps arising from such parts not having been closely specialized to any particular function, so that their modifications have not been closely checked by natural selection. It is probably from this same cause that organic beings low in the scale of nature are more variable than those which have their whole organisation more specialized, and are higher in the scale. Rudimentary organs, from being useless, will be disregarded by natural selection, and hence probably are variable. Specific characters that is, the characters which have come to differ since the several species of the same genus branched off from a common parent are more variable than generic characters, or those which have long been inherited, and have not differed within this same period. [. . .] Secondary sexual characters are highly variable, and such characters differ much in the species of the same group. Variability in the same parts of the organisation has generally been taken advantage of in giving secondary sexual differences to the sexes of the same species, and specific differences to the several species of the same genus. Any part or organ developed to an extraordinary size or in an extraordinary manner, in comparison with the same part or organ in the allied species, must have gone through an extraordinary amount of modification since the genus arose; and thus we can understand why it should often still be variable in a much higher degree than other parts; for variation is a long-continued and slow process, and natural selection will in such cases not as yet have had time to overcome the tendency to further variability and to reversion to a less modified state. But when a species with any extraordinarily-developed organ has become the parent of many modified descendants which on my view must be a very slow process, requiring a long lapse of time in this case, natural selection may readily have succeeded in giving a fixed character to the organ, in however extraordinary a manner it may be developed. Species inheriting nearly the same constitution from a common parent and exposed to similar influences will naturally tend to present analogous variations, and these same species may occasionally revert to some of the characters of their ancient progenitors. Although new and important modifications may not arise from reversion and analogous variation, such modifications will add to the beautiful and harmonious diversity of nature.

Whatever the cause may be of each slight difference in the offspring from their parents and a cause for each must exist it is the steady accumulation, through natural selection, of such differences, when beneficial to the individual, that gives rise to all the more important modifications of structure, by which the innumerable beings on the face of this earth are enabled to struggle with each other, and the best adapted to survive.

CHAPTER VI

DIFFICULTIES ON THEORY

> **Difficulties on the theory of descent with modification** – Transitions – **Absence or rarity of transitional varieties** – Transitions in habits of life – Diversified habits in the same species – Species with habits widely different from those of their allies – **Organs of extreme perfection** – Means of transition – Cases of difficulty – Natura non facit saltum – Organs of small importance – Organs not in all cases absolutely perfect – The law of Unity of Type and of the Conditions of Existence embraced by the theory of Natural Selection

*[**Game Note:** In this characteristically self-critical chapter, Darwin considers objections to his theory in detail. Reproduced here is only his general overview: Fellows are encouraged to peruse the full text of this chapter on-line. Pay particular attention to his strategy for resolving (dismissing?) apparent difficulties.]*

Long before having arrived at this part of my work, a crowd of difficulties will have occurred to the reader. Some of them are so grave that to this day I can never reflect on them without being staggered; but, to the best of my judgment, the greater number are only apparent, and those that are real are not, I think, fatal to my theory.

These difficulties and objections may be classed under the following heads: Firstly, why, if species have descended from other species by insensibly fine gradations, do we not everywhere see innumerable transitional forms? Why is not all nature in confusion instead of the species being, as we see them, well defined?

Secondly, is it possible that an animal having, for instance, the structure and habits of a bat, could have been formed by the modification of some animal with wholly different habits? Can we believe that natural selection could produce, on the one hand, organs of trifling importance, such as the tail of a giraffe, which serves as a fly-flapper, and, on the other hand, organs of such wonderful structure, as the eye, of which we hardly as yet fully understand the inimitable perfection?

Charles Darwin, the Copley Medal, and the Rise of Naturalism

Thirdly, can instincts be acquired and modified through natural selection? What shall we say to so marvelous an instinct as that which leads the bee to make cells, which have practically anticipated the discoveries of profound mathematicians?

Fourthly, how can we account for species, when crossed, being sterile and producing sterile offspring, whereas, when varieties are crossed, their fertility is unimpaired?

The two first heads shall be here discussed (with) Instinct and Hybridism in separate chapters.

[. . .] as by this theory innumerable transitional forms must have existed, why do we not find them embedded in countless numbers in the crust of the earth? It will be much more convenient to discuss this question in the chapter on the Imperfection of the geological record; and I will here only state that I believe the answer mainly lies in the record being incomparably less perfect than is generally supposed; the imperfection of the record being chiefly due to organic beings not inhabiting profound depths of the sea, and to their remains being embedded and preserved to a future age only in masses of sediment sufficiently thick and extensive to withstand an enormous amount of future degradation; and such fossiliferous masses can be accumulated only where much sediment is deposited on the shallow bed of the sea, whilst it slowly subsides. These contingencies will concur only rarely, and after enormously long intervals. Whilst the bed of the sea is stationary or is rising, or when very little sediment is being deposited, there will be blanks in our geological history. The crust of the earth is a vast museum; but the natural collections have been made only at intervals of time immensely remote.

[. . .] Organs of extreme perfection and complication. To suppose that the eye, with all its inimitable contrivances for adjusting the focus to different distances, for admitting different amounts of light, and for the correction of spherical and chromatic aberration, could have been formed by natural selection, seems, I freely confess, absurd in the highest possible degree. Yet reason tells me, that if numerous gradations from a perfect and complex eye to one very imperfect and simple, each grade being useful to its possessor, can be shown to exist; if further, the eye does vary ever so slightly, and the variations be inherited, which is certainly the case; and if any variation or modification in the organ be ever useful to an animal under changing conditions of life, then the difficulty of believing that a perfect and complex eye could be formed by natural selection, though insuperable by our imagination, can hardly be considered real. How a nerve comes to be sensitive to light, hardly concerns us more than how life itself first originated; but I may remark that several facts make me suspect that any sensitive nerve may be rendered sensitive to light, and likewise to those coarser vibrations of the air which produce sound.

CHAPTER VII

INSTINCT

> Instincts comparable with habits, but different in their origin – Instincts graduated – Aphides and ants – Instincts variable – Domestic instincts, their origin – Natural instincts of the cuckoo, ostrich, and parasitic bees – Slave-making ants – Hive-bee, its cell-making instinct – Difficulties on the theory of the Natural Selection of instincts – Neuter or sterile insects – **Summary**

*[**Game Note:** For this and several subsequent chapters I have simply reproduced Darwin's own chapter summary. While this approach greatly reduces the total amount of reading it fails to do full justice to the breadth and depth of Darwin's evidence and readers are encouraged to consult the full text concerning specific examples. **Character Alert:** The Anthropologist and the Ethnologist will want to consult the section on Slave Making Ants. As a hint to all characters – using an on-line version of the text and your browsers "Find in Page" feature will allow you to rapidly find Darwin's views on specific topics.]*

I have endeavoured briefly in this chapter to show that the mental qualities of our domestic animals vary, and that the variations are inherited. Still more briefly I have attempted to show that instincts vary slightly in a state of nature. No one will dispute that instincts are of the highest importance to each animal. Therefore I can see no difficulty, under changing conditions of life, in natural selection accumulating slight modifications of instinct to any extent, in any useful direction. In some cases habit or use and disuse have probably come into play. I do not pretend that the facts given in this chapter strengthen in any great degree my theory; but none of the cases of difficulty, to the best of my judgment, annihilate it.[27] On the other hand, the fact that instincts are not always absolutely perfect and are liable to mistakes; that no instinct has been produced for the exclusive good of other animals, but that each animal takes advantage of the instincts of others; that the canon in natural history, of "natura non facit saltum" is applicable to instincts as well as to corporeal structure, and is plainly explicable on the foregoing views, but is otherwise inexplicable, all tend to corroborate the theory of natural selection.

This theory is, also, strengthened by some few other facts in regard to instincts; as by that common case of closely allied, but certainly distinct, species, when inhabiting distant parts of the world and living under considerably different conditions of life, yet often retaining nearly the same instincts. For instance, we can understand on the principle of inheritance, how it is that the thrush of South America lines its nest with mud, in the same peculiar manner as does our British thrush: how

---

27 Darwin appreciation of the distinction between confirmation and falsification anticipates the work of Karl Popper and his claim that science is ultimately a process of "conjectures and refutations."

it is that the male wrens (Troglodytes) of North America, build "cock-nests," to roost in, like the males of our distinct Kitty-wrens, a habit wholly unlike that of any other known bird. Finally, it may not be a logical deduction, but to my imagination it is far more satisfactory to look at such instincts as the young cuckoo ejecting its foster-brothers, ants making slaves, [. . .] not as specially endowed or created instincts, but as small consequences of one general law, leading to the advancement of all organic beings, namely, multiply, vary, let the strongest live and the weakest die.

CHAPTER VIII

HYBRIDISM

> Distinction between the sterility of first crosses and of hybrids – Sterility various in degree, not universal, affected by close interbreeding, removed by domestication – Laws governing the sterility of hybrids – Sterility not a special endowment, but incidental on other differences – Causes of the sterility of first crosses and of hybrids – Parallelism between the effects of changed conditions of life and crossing – Fertility of varieties when crossed and of their mongrel offspring not universal – Hybrids and mongrels compared independently of their fertility – **Summary**

[Game Advisory: In this chapter Darwin is struggling with the implications of the fact that species, which seems so obviously distinctive, have no fixed essence. This sense of impermanence was perhaps the most disconcerting aspect of Darwin's work for his contemporaries.] First crosses between forms sufficiently distinct to be ranked as species, and their hybrids, are very generally, but not universally, sterile. The sterility is of all degrees, and is often so slight that the two most careful experimentalists who have ever lived, have come to diametrically opposite conclusions in ranking forms by this test. The sterility is innately variable in individuals of the same species, and is eminently susceptible of favourable and unfavourable conditions. The degree of sterility does not strictly follow systematic affinity, but is governed by several curious and complex laws. It is generally different, and sometimes widely different, in reciprocal crosses between the same two species. It is not always equal in degree in a first cross and in the hybrid produced from this cross.

In the same manner as in grafting trees, the capacity of one species or variety to take on another, is incidental on generally unknown differences in their vegetative systems, so in crossing, the greater or less facility of one species to unite with another, is incidental on unknown differences in their reproductive systems. There is no more reason to think that species have been specially endowed with various degrees of sterility to prevent them crossing and blending in nature, than to think that trees have been specially endowed with various and somewhat analogous degrees of difficulty in being grafted together . . .

The sterility of first crosses between pure species, which have their reproductive systems perfect, seems to depend on several circumstances; in some cases largely on the early death of the embryo. The sterility of hybrids, which have their reproductive systems imperfect, and which have had this system and their whole organisation disturbed by being compounded of two distinct species, seems closely allied to that sterility which so frequently affects pure species, when their natural conditions of life have been disturbed. This view is supported by a parallelism of another kind; namely, that the crossing of forms only slightly different is favourable to the vigour and fertility of their offspring; and that slight changes in the conditions of life are apparently favourable to the vigour and fertility of all organic beings. It is not surprising that the degree of difficulty in uniting two species, and the degree of sterility of their hybrid-offspring should generally correspond, though due to distinct causes; for both depend on the amount of difference of some kind between the species which are crossed. Nor is it surprising that the facility of effecting a first cross, the fertility of the hybrids produced, and the capacity of being grafted together though this latter capacity evidently depends on widely different circumstances should all run, to a certain extent, parallel with the systematic affinity of the forms which are subjected to experiment; for systematic affinity attempts to express all kinds of resemblance between all species.

First crosses between forms known to be varieties, or sufficiently alike to be considered as varieties, and their mongrel offspring, are very generally, but not quite universally, fertile. Nor is this nearly general and perfect fertility surprising, when we remember how liable we are to argue in a circle with respect to varieties in a state of nature; and when we remember that the greater number of varieties have been produced under domestication by the selection of mere external differences, and not of differences in the reproductive system. In all other respects, excluding fertility, there is a close general resemblance between hybrids and mongrels. Finally, then, the facts briefly given in this chapter do not seem to me opposed to, but even rather to support the view, that there is no fundamental distinction between species and varieties.

CHAPTER IX

ON THE IMPERFECTIONS OF THE GEOLOGICAL RECORD

On the absence of intermediate varieties at the present day - On the nature of extinct intermediate varieties; on their number - On the vast lapse of time, as inferred from the rate of deposition and of denudation - On the poorness of our palaeontological collections - On the intermittence of geological formations - **On the absence of intermediate varieties in any one formation - On their sudden appearance in the lowest known fossiliferous strata**

[Character Alert: Geologist, Paleontologist, Owen, Huxley—each of you should be conversant with what was known of the fossil record in the 1860s—you need to serve as authorities on such matters in both General and Council Sessions.] In the first place it should always be borne in mind what sort of intermediate forms must, on my theory, have formerly existed. I have found it difficult, when looking at any two species, to avoid picturing to myself, forms directly intermediate between them. But this is a wholly false view; we should always look for forms intermediate between each species and a common but unknown progenitor; and the progenitor will generally have differed in some respects from all its modified descendants. To give a simple illustration: the fantail and pouter pigeons have both descended from the rock-pigeon; if we possessed all the intermediate varieties which have ever existed, we should have an extremely close series between both and the rock-pigeon; but we should have no varieties directly intermediate between the fantail and pouter; none, for instance, combining a tail somewhat expanded with a crop somewhat enlarged, the characteristic features of these two breeds. These two breeds, moreover, have become so much modified, that if we had no historical or indirect evidence regarding their origin, it would not have been possible to have determined from a mere comparison of their structure with that of the rock-pigeon, whether they had descended from this species or from some other allied species, such as C. oenas.

So with natural species, if we look to forms very distinct, for instance to the horse and tapir, we have no reason to suppose that links ever existed directly intermediate between them, but between each and an unknown common parent. The common parent will have had in its whole organisation much general resemblance to the tapir and to the horse; but in some points of structure may have differed considerably from both, even perhaps more than they differ from each other. Hence in all such cases, we should be unable to recognise the parent-form of any two or more species, even if we closely compared the structure of the parent with that of its modified descendants, unless at the same time we had a nearly perfect chain of the intermediate links.

[. . .] The abrupt manner in which whole groups of species suddenly appear in certain formations, has been urged by several palaeontologists, for instance, by Agassiz, Pictet, and by none more forcibly than by Professor Sedgwick, as a fatal objection to the belief in the transmutation of species. If numerous species, belonging to the same genera or families, have really started into life all at once, the fact would be fatal to the theory of descent with slow modification through natural selection. For the development of a group of forms, all of which have descended from some one progenitor, must have been an extremely slow process; and the progenitors must have lived long ages before their modified descendants. But we continually over-rate the perfection of the geological record, and falsely infer, because certain genera or families have not been found beneath a certain stage, that they did not exist before that stage. We continually forget how large the world is, compared with the area over which our geological formations have been carefully examined; we forget that groups of species may elsewhere have long existed and have slowly multiplied before they invaded the ancient archipelagoes of Europe and of the United States. We do not make due allowance for the enormous intervals of time, which have probably elapsed between our consecutive formations, longer perhaps in some cases than the time required for the accumulation of each formation. These intervals will have given time for the multiplication of species from some one or some few parent-forms; and in the succeeding formation such species will appear as if suddenly created.

CHAPTER X

ON THE GEOLOGICAL SUCCESSION OF ORGANIC BEINGS

On the slow and successive appearance of new species - On their different rates of change - Species once lost do not reappear - Groups of species follow the same general rules in their appearance and disappearance as do single species - On Extinction - On simultaneous changes in the forms of life throughout the world - On the affinities of extinct species to each other and to living species - On the state of development of ancient forms - On the succession of the same types within the same areas - *Summary of preceding and present chapters*

*[**Game Note:** Absence of evidence is not evidence of absence. X-Men can interpret Darwin's theory as a guide to interesting empirical studies. A-Men can make the case the sparsity of evidence points to the recklessly speculative nature of Darwin's work.]* I have attempted to show that the geological record is extremely imperfect; that only a small portion of the globe has been geologically explored with care; that only certain classes of organic beings have been largely preserved in a fossil state; that the number both of specimens and of species, preserved in our museums, is absolutely as nothing compared with the incalculable number of generations which must have passed away even during a single formation; that, owing to subsidence being necessary for the accumulation of fossiliferous deposits thick enough to resist future degradation, enormous intervals of time have elapsed between the successive formations; that there has probably been more extinction during the periods of subsidence, and more variation during the periods of elevation, and during the latter the record will have been least perfectly kept; that each single formation has not been continuously deposited; that the duration of each formation is, perhaps, short compared with the average duration of specific forms; that migration has played an important part in the first appearance of new forms in any one area and formation; that widely ranging species are those which have varied most, and have oftenest given rise to new species; and that varieties have at first often been local. All these causes taken conjointly, must have tended to make the geological record extremely imperfect, and will to a large extent explain why we do not find

interminable varieties, connecting together all the extinct and existing forms of life by the finest graduated steps.

He who rejects these views on the nature of the geological record, will rightly reject my whole theory. For he may ask in vain where are the numberless transitional links which must formerly have connected the closely allied or representative species, found in the several stages of the same great formation. He may disbelieve in the enormous intervals of time which have elapsed between our consecutive formations; he may overlook how important a part migration must have played, when the formations of any one great region alone, as that of Europe, are considered; he may urge the apparent, but often falsely apparent, sudden coming in of whole groups of species. He may ask where are the remains of those infinitely numerous organisms which must have existed long before the first bed of the Silurian system was deposited: I can answer this latter question only hypothetically, by saying that as far as we can see, where our oceans now extend they have for an enormous period extended, and where our oscillating continents now stand they have stood ever since the Silurian epoch; but that long before that period, the world may have presented a wholly different aspect; and that the older continents, formed of formations older than any known to us, may now all be in a metamorphosed condition, or may lie buried under the ocean.[28]

Passing from these difficulties, all the other great leading facts in palaeontology seem to me simply to follow on the theory of descent with modification through natural selection. We can thus understand how it is that new species come in slowly and successively; how species of different classes do not necessarily change together, or at the same rate, or in the same degree; yet in the long run that all undergo modification to some extent. The extinction of old forms is the almost inevitable consequence of the production of new forms.[29] [. . .] Groups of species increase in numbers slowly, and endure for unequal periods of time; for the process of modification is necessarily slow, and depends on many complex contingencies. The dominant species of the larger dominant groups tend to leave many modified descendants, and thus new sub-groups and groups are formed. As these are formed, the species of the less vigorous groups, from their inferiority inherited from a common progenitor, tend to become extinct together, and to leave no modified offspring on the face of the earth. But the utter extinction of a whole group of species may

28 Numerous pre-Silurian fossil beds have since been discovered. S.J. Gould [(1989) Wonderful life: the Burgess Shale and the nature of history. W.W. Norton: New York NY] provides a particularly rich and accessible account of the diversity of one of the best studied pre-Silurian (actually pre-Cambrian, 530 m.y.a.) sites.

29 From a purely logical perspective, it makes at least as much sense to argue that the production of new forms is the almost inevitable consequence of the opportunity provided by extinction of old forms; this is especially important in light of relatively recent realization that there have been at least seven great extinction events in earth's history (independent of species interactions!). See http://en.wikipedia.org/wiki/Extinction_event for more details.

often be a very slow process, from the survival of a few descendants, lingering in protected and isolated situations. When a group has once wholly disappeared, it does not reappear; for the link of generation has been broken.

We can understand how the spreading of the dominant forms of life, which are those that oftenest vary, will in the long run tend to people the world with allied, but modified, descendants; and these will generally succeed in taking the places of those groups of species which are their inferiors in the struggle for existence. Hence, after long intervals of time, the productions of the world will appear to have changed simultaneously. [. . .]

The inhabitants of each successive period in the world's history have beaten their predecessors in the race for life, and are, in so far, higher in the scale of nature; and this may account for that vague yet ill-defined sentiment, felt by many palaeontologists, that organisation on the whole has progressed. If it should hereafter be proved that ancient animals resemble to a certain extent the embryos of more recent animals of the same class, the fact will be intelligible. The succession of the same types of structure within the same areas during the later geological periods ceases to be mysterious, and is simply explained by inheritance.

If the geological record be as imperfect as I believe it to be, and it may at least be asserted that the record cannot be proved to be much more perfect, the main objections to the theory of natural selection are greatly diminished or disappear. On the other hand, all the chief laws of palaeontology plainly proclaim, as it seems to me, that species have been produced by ordinary generation: old forms having been supplanted by new and improved forms of life, produced by the laws of variation still acting round us, and preserved by Natural Selection.

CHAPTER XI

GEOGRAPHICAL DISTRIBUTION

Present distribution cannot be accounted for by differences in physical conditions - Importance of barriers - Affinity of the productions of the same continent - Centres of creation - **Means of dispersal, by changes of climate and of the level of the land, and by occasional means** - Dispersal during the Glacial period co-extensive with the world

[Game Note: Recall that, like Darwin, many of you have traveled rather extensively and have been personally struck by the patterns he is seeking to interpret here. Depending on your objectives, you may elect to confirm or contrast his observations with your own.] In considering the distribution of organic beings over the face of the globe, the first great fact which strikes us is, that neither the similarity

nor the dissimilarity of the inhabitants of various regions can be accounted for by their climatal and other physical conditions. Of late, almost every author who has studied the subject has come to this conclusion. The case of America alone would almost suffice to prove its truth: for if we exclude the northern parts where the circumpolar land is almost continuous, all authors agree that one of the most fundamental divisions in geographical distribution is that between the New and Old Worlds;[30] yet if we travel over the vast American continent, from the central parts of the United States to its extreme southern point, we meet with the most diversified conditions; the most humid districts, arid deserts, lofty mountains, grassy plains, forests, marshes, lakes, and great rivers, under almost every temperature. There is hardly a climate or condition in the Old World which cannot be paralleled in the New at least as closely as the same species generally require; for it is a most rare case to find a group of organisms confined to any small spot, having conditions peculiar in only a slight degree; for instance, small areas in the Old World could be pointed out hotter than any in the New World, yet these are not inhabited by a peculiar fauna or flora. Notwithstanding this parallelism in the conditions of the Old and New Worlds, how widely different are their living productions!

In the southern hemisphere, if we compare large tracts of land in Australia, South Africa, and western South America, between latitudes 25° and 35°, we shall find parts extremely similar in all their conditions, yet it would not be possible to point out three faunas and floras more utterly dissimilar. Or again we may compare the productions of South America south of lat. 35° with those north of 25°, which consequently inhabit a considerably different climate, and they will be found incomparably more closely related to each other, than they are to the productions of Australia or Africa under nearly the same climate. Analogous facts could be given with respect to the inhabitants of the sea.

A second great fact which strikes us in our general review is, that barriers of any kind, or obstacles to free migration, are related in a close and important manner to the differences between the productions of various regions. We see this in the great difference of nearly all the terrestrial productions of the New and Old Worlds, excepting in the northern parts, where the land almost joins, and where, under a slightly different climate, there might have been free migration for the northern temperate forms, as there now is for the strictly arctic productions. We see the same fact in the great difference between the inhabitants of Australia, Africa, and South America under the same latitude: for these countries are almost as much isolated from each other as is possible. On each continent, also, we see the same fact; for on the opposite sides of lofty and continuous mountain-ranges, and of great deserts, and sometimes even of large rivers, we find different productions; though as moun-

---

30 Remember that Darwin's work here far pre-dates the notion of continental drift which was first clearly articulated by Alfred Wegener in 1912 and later incorporated into the larger theory of plate tectonics.

tain chains, deserts, &c., are not as impassable, or likely to have endured so long as the oceans separating continents, the differences are very inferior in degree to those characteristic of distinct continents.[31]

[. . .] Considering that the several above means of transport, and that several other means, which without doubt remain to be discovered, have been in action year after year, for centuries and tens of thousands of years, it would I think be a marvellous fact if many plants had not thus become widely transported. These means of transport are sometimes called accidental, but this is not strictly correct: the currents of the sea are not accidental, nor is the direction of prevalent gales of wind. It should be observed that scarcely any means of transport would carry seeds for very great distances; for seeds do not retain their vitality when exposed for a great length of time to the action of seawater; nor could they be long carried in the crops or intestines of birds. These means, however, would suffice for occasional transport across tracts of sea some hundred miles in breadth, or from island to island, or from a continent to a neighbouring island, but not from one distant continent to another. The floras of distant continents would not by such means become mingled in any great degree; but would remain as distinct as we now see them to be. The currents, from their course, would never bring seeds from North America to Britain, though they might and do bring seeds from the West Indies to our western shores, where, if not killed by so long an immersion in salt-water, they could not endure our climate. Almost every year, one or two land-birds are blown across the whole Atlantic Ocean, from North America to the western shores of Ireland and England; but seeds could be transported by these wanderers only by one means, namely, in dirt sticking to their feet, which is in itself a rare accident. Even in this case, how small would the chance be of a seed falling on favourable soil, and coming to maturity! But it would be a great error to argue that because a well-stocked island, like Great Britain, has not, as far as is known (and it would be very difficult to prove this), received within the last few centuries, through occasional means of transport, immigrants from Europe or any other continent, that a poorly-stocked island, though standing more remote from the mainland, would not receive colonists by similar means. I do not doubt that out of twenty seeds or animals transported to an island, even if far less well-stocked than Britain, scarcely more than one would be so well fitted to its new home, as to become naturalised. But this, as it seems to me, is no valid argument against what would be effected by occasional means of transport, during the long lapse of geological time, whilst an island was being upheaved and formed, and before it had become fully stocked with inhabitants. On almost bare land, with few or no destructive insects or birds living there, nearly every seed, which chanced to arrive, would be sure to germinate and survive.

31 There has been considerable work on the relative significance of migration versus separation of populations by geological events. Because earth's crust has proven far more dynamic than Darwin supposed, he generally over-estimated the significance of migration and underestimated the role of vicariance events (e.g., emergence of the isthmus of Panama, plate tectonics more generally).

Charles Darwin, the Copley Medal, and the Rise of Naturalism

CHAPTER XII

GEOGRAPHICAL DISTRIBUTION (CONTINUED)

Distribution of fresh-water productions - On the inhabitants of oceanic islands - Absence of Batrachians and of terrestrial Mammals - On the relations of the inhabitants of islands to those of the nearest mainland - On colonisation from the nearest source with subsequent modification - **Summary of the last and present chapters**

*[Game Note: Darwin is showing here that his theory produces what Whewell championed as a "consilience of inductions"—an ability of a theory to make coherent sense of an other large class of disparate facts. **Character Alert:** Inductivist (Baconian) Philosopher and Friend of Mill.]* In these chapters I have endeavoured to show, that if we make due allowance for our ignorance of the full effects of all the changes of climate and of the level of the land, which have certainly occurred within the recent period, and of other similar changes which may have occurred within the same period; if we remember how profoundly ignorant we are with respect to the many and curious means of occasional transport, a subject which has hardly ever been properly experimentised on; if we bear in mind how often a species may have ranged continuously over a wide area, and then have become extinct in the intermediate tracts, I think the difficulties in believing that all the individuals of the same species, wherever located, have descended from the same parents, are not insuperable. And we are led to this conclusion, which has been arrived at by many naturalists under the designation of single centres of creation, by some general considerations, more especially from the importance of barriers and from the analogical distribution of sub-genera, genera, and families.

If the difficulties be not insuperable in admitting that in the long course of time the individuals of the same species, and likewise of allied species, have proceeded from some one source; then I think all the grand leading facts of geographical distribution are explicable on the theory of migration (generally of the more dominant forms of life), together with subsequent modification and the multiplication of new forms. We can thus understand the high importance of barriers, whether of land or water, which separate our several zoological and botanical provinces. We can thus understand the localisation of sub-genera, genera, and families; and how it is that under different latitudes, for instance in South America, the inhabitants of the plains and mountains, of the forests, marshes, and deserts, are in so mysterious a manner linked together by affinity, and are likewise linked to the extinct beings which formerly inhabited the same continent. Bearing in mind that the mutual relations of organism to organism are of the highest importance, we can see why two areas having nearly the same physical conditions should often be inhabited by very different forms of life; for according to the length of time which has elapsed since new inhabitants entered one region; according to the nature of the communication which allowed certain forms and not others to enter, either in greater or lesser numbers; according or not, as those which entered happened to come in more or less direct competition with each other and with the aborigines; and according as the immigrants were capable of varying more or less rapidly, there would ensue in different regions, independently of their physical conditions, infinitely diversified conditions of life, there would be an almost endless amount of organic action and reaction, and we should find, as we do find, some groups of beings greatly, and some only slightly modified, some developed in great force, some existing in scanty numbers in the different great geographical provinces of the world.

On these same principles, we can understand, as I have endeavoured to show, why oceanic islands should have few inhabitants, but of these a great number should be endemic or peculiar; and why, in relation to the means of migration, one group of beings, even within the same class, should have all its species endemic, and another group should have all its species common to other quarters of the world. We can see why whole groups of organisms, as batrachians and terrestrial mammals, should be absent from oceanic islands, whilst the most isolated islands possess their own peculiar species of aërial mammals or bats. We can see why there should be some relation between the presence of mammals, in a more or less modified condition, and the depth of the sea between an island and the mainland. We can clearly see why all the inhabitants of an archipelago, though specifically distinct on the several islets, should be closely related to each other, and likewise be related, but less closely, to those of the nearest continent or other source whence immigrants were probably derived. We can see why in two areas, however distant from each other, there should be a correlation, in the presence of identical species, of varieties, of doubtful species, and of distinct but representative species.

[. . .] On my theory these several relations throughout time and space are intelligible; for whether we look to the forms of life which have changed during successive ages within the same quarter of the world, or to those which have changed after having migrated into distant quarters, in both cases the forms within each class have been connected by the same bond of ordinary generation; and the more nearly any two forms are related in blood, the nearer they will generally stand to each other in time and space; in both cases the laws of variation have been the same, and modifications have been accumulated by the same power of natural selection.

Charles Darwin, the Copley Medal, and the Rise of Naturalism

CHAPTER XIII

MUTUAL AFFINITIES OF ORGANIC BEINGS: MORPHOLOGY: EMBRYOLOGY: RUDIMENTARY ORGANS

CLASSIFICATION, groups subordinate to groups - Natural system - Rules and difficulties in classification, explained on the theory of descent with modification - Classification of varieties - Descent always used in classification - Analogical or adaptive characters - Affinities, general, complex and radiating - Extinction separates and defines groups - MORPHOLOGY, between members of the same class, between parts of the same individual - EMBRYOLOGY, laws of, explained by variations not supervening at an early age, and being inherited at a corresponding age - RUDIMENTARY ORGANS; their origin explained - **Summary**

In this chapter I have attempted to show, that the subordination of group to group in all organisms throughout all time; that the nature of the relationship, by which all living and extinct beings are united by complex, radiating, and circuitous lines of affinities into one grand system; the rules followed and the difficulties encountered by naturalists in their classifications; the value set upon characters, if constant and prevalent, whether of high vital importance, or of the most trifling importance, or, as in rudimentary organs, of no importance; the wide opposition in value between analogical or adaptive characters, and characters of true affinity; and other such rules; all naturally follow on the view of the common parentage of those forms which are considered by naturalists as allied, together with their modification through natural selection, with its contingencies of extinction and divergence of character. In considering this view of classification, it should be borne in mind that the element of descent has been universally used in ranking together the sexes, ages, and acknowledged varieties of the same species, however different they may be in structure. If we extend the use of this element of descent, the only certainly known cause of similarity in organic beings, we shall understand what is meant by the natural system: it is genealogical in its attempted arrangement, with the grades of acquired difference marked by the terms varieties, species, genera, families, orders, and classes.

On this same view of descent with modification, all the great facts in Morphology become intelligible, whether we look to the same pattern displayed in the homologous organs, to whatever purpose applied, of the different species of a class; or to homologous parts constructed on the same pattern in each individual animal and plant.

On the principle of successive slight variations, not necessarily or generally supervening at a very early period of life, and being inherited at a corresponding period, we can understand the great leading facts in Embryology; namely, the resemblance in an individual embryo of the homologous parts, which when matured will become widely different from each other in structure and function; and the resemblance in different species of a class of the homologous parts or organs, though fitted in the adult members for purposes as different as possible. Larvae are active embryos, which have become specially modified in relation to their habits of life, through the principle of modifications being inherited at corresponding ages. On this same principle and bearing in mind, that when organs are reduced in size, either from disuse or selection, it will generally be at that period of life when the being has to provide for its own wants, and bearing in mind how strong is the principle of inheritance the occurrence of rudimentary organs and their final abortion, present to us no inexplicable difficulties; on the contrary, their presence might have been even anticipated. The importance of embryological characters and of rudimentary organs in classification is intelligible, on the view that an arrangement is only so far natural as it is genealogical.[32]

[**Closing paragraph added to later editions:** Finally, the several classes of facts which have been considered in this chapter, seem to me to proclaim so plainly, that the innumerable species, genera and families, with which this world is peopled, are all descended , each within its own class or group, from common parents, and have all been modified in the course of descent, that I should without hesitation adopt this view, even if it were unsupported by other arguments.][33]

CHAPTER XIV

RECAPITULATION AND CONCLUSION

Recapitulation of the difficulties on the theory of Natural Selection – Recapitulation of the general and special circumstances in its favour – Causes of the general belief in the immutability of species – How far the theory of natural selection may be extended – Effects of its adoption on the study of Natural history – Concluding remarks

[*Game Note: Excerpts presented here are drawn primarily from Darwin's concluding remarks; readers may wish to read the entire final chapter on-line for Darwin's full recapitulation of his "long argument."*]

---

32 For an update on the relationship between evolution and development see J. Gerhart and M. Kirshner [(1997) Cells, embryos, and evolution: toward a cellular and developmental understanding of phenotypic variation and evolutionary adaptability. Blackwell Science: Malden MA.]

33 This addition clearly reflects a Darwin's commitment to Whewell's "consilience of inductions." It is also interesting to note that Darwin here explicitly restricts his theory to changes within group or class.

Charles Darwin, the Copley Medal, and the Rise of Naturalism

As this whole volume is one long argument, it may be convenient to the reader to have the leading facts and inferences briefly recapitulated. [. . .] That many and grave objections may be advanced against the theory of descent with modification through natural selection, I do not deny. I have endeavoured to give to them their full force. Nothing at first can appear more difficult to believe than that the more complex organs and instincts should have been perfected not by means superior to, though analogous with, human reason, but by the accumulation of innumerable slight variations, each good for the individual possessor. Nevertheless, this difficulty, though appearing to our imagination insuperably great, cannot be considered real if we admit the following propositions, namely,—that gradations in the perfection of any organ or instinct, which we may consider, either do now exist or could have existed, each good of its kind,—that all organs and instincts are, in ever so slight a degree, variable,—and, lastly, that there is a struggle for existence leading to the preservation of each profitable deviation of structure or instinct. The truth of these propositions cannot, I think, be disputed. [. . .]

[. . .] As natural selection acts solely by accumulating slight, successive, favourable variations, it can produce no great or sudden modification; it can act only by very short and slow steps. Hence the canon of "Natura non facit saltum," which every fresh addition to our knowledge tends to make more strictly correct, is on this theory simply intelligible. We can plainly see why nature is prodigal in variety, though niggard in innovation. But why this should be a law of nature if each species has been independently created, no man can explain.[34]

Many other facts are, as it seems to me, explicable on this theory. How strange it is that a bird, under the form of woodpecker, should have been created to prey on insects on the ground; that upland geese, which never or rarely swim, should have been created with webbed feet; that a thrush should have been created to dive and feed on sub-aquatic insects; and that a petrel should have been created with habits and structure fitting it for the life of an auk or grebe! and so on in endless other cases. But on the view of each species constantly trying to increase in number, with natural selection always ready to adapt the slowly varying descendants of each to any unoccupied or ill-occupied place in nature, these facts cease to be strange, or perhaps might even have been anticipated.

As natural selection acts by competition, it adapts the inhabitants of each country only in relation to the degree of perfection of their associates; so that we need feel no surprise at the inhabitants of any one country, although on the ordinary view supposed to have been specially created and adapted for that country, being beaten

34 See N. Eldredge and S.J. Gould [(1972) Punctuated equilibrium: an alternative to phyletic gradualism. In: Schopf, T. Models in Paleobiology. W.H. Freeman: San Francisco CA] for a discussion of how gradual changes in genes can produce more abrupt changes in form. The on-going exchange between gradualists and proponents of punctuated equilibrium continues to be one of the most dynamic and contentious arenas of evolutionary theory.

and supplanted by the naturalised productions from another land. *[Naturalizing and justifying the British Empire?]* Nor ought we to marvel if all the contrivances in nature be not, as far as we can judge, absolutely perfect; and if some of them be abhorrent to our ideas of fitness. We need not marvel at the sting of the bee causing the bee's own death; at drones being produced in such vast numbers for one single act, and being then slaughtered by their sterile sisters; at the astonishing waste of pollen by our fir-trees; at the instinctive hatred of the queen bee for her own fertile daughters; at ichneumonidae feeding within the live bodies of caterpillars; and at other such cases. The wonder indeed is, on the theory of natural selection, that more cases of the want of absolute perfection have not been observed.

The fact, as we have seen, that all past and present organic beings constitute one grand natural system, with group subordinate to group, and with extinct groups often falling in between recent groups, is intelligible on the theory of natural selection with its contingencies of extinction and divergence of character.[35] On these same principles we see how it is, that the mutual affinities of the species and genera within each class are so complex and circuitous. We see why certain characters are far more serviceable than others for classification;—why adaptive characters, though of paramount importance to the being, are of hardly any importance in classification; why characters derived from rudimentary parts, though of no service to the being, are often of high classificatory value; and why embryological characters are the most valuable of all. The real affinities of all organic beings are due to inheritance or community of descent. The natural system is a genealogical arrangement, in which we have to discover the lines of descent by the most permanent characters, however slight their vital importance may be.

The framework of bones being the same in the hand of a man, wing of a bat, fin of the porpoise, and leg of the horse,—the same number of vertebrae forming the neck of the giraffe and of the elephant,—and innumerable other such facts, at once explain themselves on the theory of descent with slow and slight successive modifications. The similarity of pattern in the wing and leg of a bat, though used for such different purposes, —in the jaws and legs of a crab,—in the petals, stamens, and pistils of a flower, is likewise intelligible on the view of the gradual modification of parts or organs, which were alike in the early progenitor of each class. On the principle of successive variations not always supervening at an early age, and being inherited at a corresponding not early period of life, we can clearly see why the embryos of mammals, birds, reptiles, and fishes should be so closely alike, and should be so unlike the adult forms. We may cease marvelling at the embryo of an air-breathing mammal or bird having branchial slits and arteries running in loops, like those in a fish which has to breathe the air dissolved in water, by the aid of well-developed branchiae.

---

35 Systematics is slowly but surely being re-built on an explicitly evolutionary or "cladistic" foundation as first advocated by W. Hennig (1966) in "Phylogenetic Systematics" [University of Illinois Press: Urbana IL].

Charles Darwin, the Copley Medal, and the Rise of Naturalism

Disuse, aided sometimes by natural selection, will often tend to reduce an organ, when it has become useless by changed habits or under changed conditions of life; and we can clearly understand on this view the meaning of rudimentary organs. But disuse and selection will generally act on each creature, when it has come to maturity and has to play its full part in the struggle for existence, and will thus have little power of acting on an organ during early life; hence the organ will not be much reduced or rendered rudimentary at this early age. The calf, for instance, has inherited teeth, which never cut through the gums of the upper jaw, from an early progenitor having well-developed teeth; and we may believe, that the teeth in the mature animal were reduced, during successive generations, by disuse or by the tongue and palate having been fitted by natural selection to browse without their aid; whereas in the calf, the teeth have been left untouched by selection or disuse, and on the principle of inheritance at corresponding ages have been inherited from a remote period to the present day. On the view of each organic being and each separate organ having been specially created, how utterly inexplicable it is that parts, like the teeth in the embryonic calf or like the shrivelled wings under the soldered wing-covers of some beetles, should thus so frequently bear the plain stamp of inutility! Nature may be said to have taken pains to reveal, by rudimentary organs and by homologous structures, her scheme of modification, which it seems that we wilfully will not understand.[36] [. . .]

[. . .] I have now recapitulated the chief facts and considerations which have thoroughly convinced me that species have changed, and are still slowly changing by the preservation and accumulation of successive slight favourable variations. Why, it may be asked, have all the most eminent living naturalists and geologists rejected this view of the mutability of species? It cannot be asserted that organic beings in a state of nature are subject to no variation; it cannot be proved that the amount of variation in the course of long ages is a limited quantity; no clear distinction has been, or can be, drawn between species and well-marked varieties. It cannot be maintained that species when intercrossed are invariably sterile, and varieties invariably fertile; or that sterility is a special endowment and sign of creation. The belief that species were immutable productions was almost unavoidable as long as the history of the world was thought to be of short duration; and now that we have acquired some idea of the lapse of time, we are too apt to assume, without proof, that the geological record is so perfect that it would have afforded us plain evidence of the mutation of species, if they had undergone mutation.

But the chief cause of our natural unwillingness to admit that one species has given birth to other and distinct species, is that we are always slow in admitting any great change of which we do not see the intermediate steps. The difficulty is the same as that felt by so many geologists, when Lyell first insisted that long lines of inland

36 For a classic and wonderfully insightful extension of these ideas see F. Jacob (1977) Evolution and tinkering. Science 196: 1161-1166.

cliffs had been formed, and great valleys excavated, by the slow action of the coast-waves. The mind cannot possibly grasp the full meaning of the term of a hundred million years; it cannot add up and perceive the full effects of many slight variations, accumulated during an almost infinite number of generations.

Although I am fully convinced of the truth of the views given in this volume under the form of an abstract, I by no means expect to convince experienced naturalists whose minds are stocked with a multitude of facts all viewed, during a long course of years, from a point of view directly opposite to mine. It is so easy to hide our ignorance under such expressions as the "plan of creation," "unity of design," &c., and to think that we give an explanation when we only restate a fact. Any one whose disposition leads him to attach more weight to unexplained difficulties than to the explanation of a certain number of facts will certainly reject my theory. A few naturalists, endowed with much flexibility of mind, and who have already begun to doubt on the immutability of species, may be influenced by this volume; but I look with confidence to the future, to young and rising naturalists, who will be able to view both sides of the question with impartiality. Whoever is led to believe that species are mutable will do good service by conscientiously expressing his conviction; for only thus can the load of prejudice by which this subject is overwhelmed be removed.

Several eminent naturalists have of late published their belief that a multitude of reputed species in each genus are not real species; but that other species are real, that is, have been independently created. This seems to me a strange conclusion to arrive at. They admit that a multitude of forms, which till lately they themselves thought were special creations, and which are still thus looked at by the majority of naturalists, and which consequently have every external characteristic feature of true species,—they admit that these have been produced by variation, but they refuse to extend the same view to other and very slightly different forms. Nevertheless they do not pretend that they can define, or even conjecture, which are the created forms of life, and which are those produced by secondary laws. They admit variation as a *vera causa* in one case, they arbitrarily reject it in another, without assigning any distinction in the two cases. The day will come when this will be given as a curious illustration of the blindness of preconceived opinion. These authors seem no more startled at a miraculous act of creation than at an ordinary birth. But do they really believe that at innumerable periods in the earth's history certain elemental atoms have been commanded suddenly to flash into living tissues? Do they believe that at each supposed act of creation one individual or many were produced? Were all the infinitely numerous kinds of animals and plants created as eggs or seed, or as full grown? and in the case of mammals, were they created bearing the false marks of nourishment from the mother's womb? Although naturalists very properly demand a full explanation of every difficulty from those who believe in the mutability of species, on their own side they ignore the whole subject of the first appearance of species in what they consider reverent silence.

It may be asked how far I extend the doctrine of the modification of species. The question is difficult to answer, because the more distinct the forms are which we may consider, by so much the arguments fall away in force. But some arguments of the greatest weight extend very far. All the members of whole classes can be connected together by chains of affinities, and all can be classified on the same principle, in groups subordinate to groups. Fossil remains sometimes tend to fill up very wide intervals between existing orders. Organs in a rudimentary condition plainly show that an early progenitor had the organ in a fully developed state; this in some instances necessarily implies an enormous amount of modification in the descendants. Throughout whole classes various structures are formed on the same pattern, and at an embryonic age the species closely resemble each other. Therefore I cannot doubt that the theory of descent with modification embraces all the members of the same class. I believe that animals have descended from at most only four or five progenitors, and plants from an equal or lesser number.

Analogy would lead me one step further, namely, to the belief that all animals and plants have descended from some one prototype. But analogy may be a deceitful guide. Nevertheless all living things have much in common, in their chemical composition, their germinal vesicles, their cellular structure, and their laws of growth and reproduction. [. . .] Therefore I should infer from analogy that probably all the organic beings which have ever lived on this earth have descended from some one primordial form, into which life was first breathed.

When the views entertained in this volume *On the Origin of Species* [. . .] are generally admitted, we can dimly foresee that there will be a considerable revolution in natural history. Systematists will be able to pursue their labours as at present; but they will not be incessantly haunted by the shadowy doubt whether this or that form be in essence a species. [. . .] Hereafter we shall be compelled to acknowledge that the only distinction between species and well-marked varieties is, that the latter are known, or believed, to be connected at the present day by intermediate gradations, whereas species were formerly thus connected. [. . .] In short, we shall have to treat species in the same manner as those naturalists treat genera, who admit that genera are merely artificial combinations made for convenience. This may not be a cheering prospect; but we shall at least be freed from the vain search for the undiscovered and undiscoverable essence of the term species.

[. . .] When we no longer look at an organic being as a savage looks at a ship, as at something wholly beyond his comprehension; when we regard every production of nature as one which has had a history; when we contemplate every complex structure and instinct as the summing up of many contrivances, each useful to the possessor, [. . .] how far more interesting, I speak from experience, will the study of natural history become! [. . .] In the distant future I see open fields for far more important researches. Psychology will be based on a new foundation, that of the necessary acquirement of each mental power and capacity by gradation. Light will be thrown on the origin of man and his history.[37]

Authors of the highest eminence seem to be fully satisfied with the view that each species has been independently created. To my mind it accords better with what we know of the laws impressed on matter by the Creator, that the production and extinction of the past and present inhabitants of the world should have been due to secondary causes, like those determining the birth and death of the individual. When I view all beings not as special creations, but as the lineal descendants of some few beings which lived long before the first bed of the Silurian system was deposited, they seem to me to become ennobled. [. . .] As all the living forms of life are the lineal descendants of those which lived long before [. . .], we may feel certain that the ordinary succession by generation has never once been broken, and that no cataclysm has desolated the whole world. Hence we may look with some confidence to a secure future of equally inappreciable length. And as natural selection works solely by and for the good of each being, all corporeal and mental endowments will tend to progress towards perfection.

It is interesting to contemplate an entangled bank, clothed with many plants of many kinds, with birds singing on the bushes, with various insects flitting about, and with worms crawling through the damp earth, and to reflect that these elaborately constructed forms, so different from each other, and dependent on each other in so complex a manner, have all been produced by laws acting around us. [. . .] There is grandeur in this view of life, with its several powers, having been originally breathed[38] into a few forms or into one; and that, whilst this planet has gone cycling on according to the fixed law of gravity, from so simple a beginning endless forms most beautiful and most wonderful have been, and are being, evolved.[39]

GLOSSARY

I am indebted to the kindness of Mr. W. S. Dallas for this Glossary, which has been given because several readers have complained to me that some of the terms used were unintelligible to them. Mr. Dallas has endeavoured to give the explanations of the terms in as popular a form as possible.

---

37 The recent emergence of evolutionary psychology marks the fulfillment of this possibility. For a concise introductory overview of evolutionary psychology, see http://en.wikipedia.org/wiki/Evolutionary_psychology.

38 Darwin added the phrase "by the creator" here in the second and all subsequent edition. He did this to "conciliate angry clerics" but as a letter of 29 March 1863 to Hooker reveals he regretted his decision: "I have long since regretted that I truckled to public opinion and used the Pentateuchal term of creation, by which I really meant 'appeared' by some wholly unknown process." In the same letter Darwin wrote, "It is mere rubbish, thinking at present of the origin of life; one might as well think of the origin of matter."

39 See P.R. Sloan [(2001) The sense of sublimity: Darwin on nature and divinity. Osiris 16: 251-269] for insightful analysis of Darwin's 'nature reveries' and his philosophical and theological affinity with Alexander von Humboldt.

Figure 3. From Daniel Girton (1790) *The Complete Pigeon Fancier.* London: Alexander Hogg

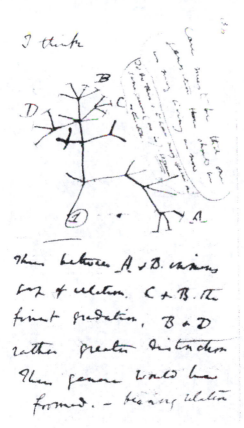

Figure 4. Darwin 1837 – A descent tree from Darwin's B Notebook (Cambridge University Library)

PEDIGREE OF MAN.

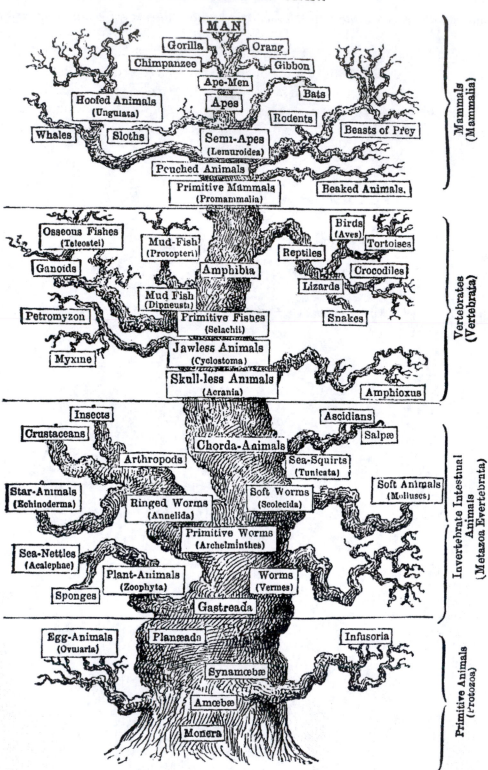

Figure 5. Ernst Haeckel's 1874 "Lebensbaum" ("Tree of Life") image (© Dorling Kindersley)

APPENDIX B. PRIMARY SOURCE DOCUMENTS

Samuel Wilberforce's review of *On the Origin of Species* is especially important to gain an understanding of the critical responses of Darwin's contemporaries. Other reviews are available for your study through internet research and library holdings. Selections from the *Darwin Correspondence Project* will demonstrate Darwin's complex relationships with his fellow members of the Royal Society and his friends and neighbors. William Paley's *Natural Theology* sets out the prevailing theological understanding of science that so influenced all the characters in this game. John Lubbock's "On Tact," will be useful to provide you with a guide to appropriate behavior for the time.

SAMUEL WILBERFORCE, REVIEW OF *ON THE ORIGIN OF SPECIES*

[Wilberforce's "Darwin's *Origin of Species*" originally appeared in *Quarterly Review*, 1860, pp. 225-264. The text below was taken from a text-fiche of Samuel Wilberforce, *Essays Contributed to the "Quarterly Review,"* London: J. Murray, 1874 and was edited and abridged for this game. Editor's notes appear in brackets [ ].]

ART. VII.—*On the Origin of Species, by means of Natural Selection; or the Preservation of Favoured Races in the Struggle for Life.* By Charles Darwin, M. A., F.R.S. London, 1860.

ANY contribution to our Natural History literature from the pen of Mr. C. Darwin is certain to command attention. His scientific attainments, his insight and carefulness as an observer, blended with no scanty measure of imaginative sagacity, and his clear and lively style, make all his writings unusually attractive. His present volume *On the Origin of Species* is the result of many years of observation, thought, and speculation; and is manifestly regarded by him as the "opus" upon which his future fame is to rest. It is true that he announces it modestly enough as the mere precursor of a mightier volume. But that volume is only intended to supply the facts which are to support the completed argument of the present essay. In this we have a specimen-collection of the vast accumulation; and, working from these as the high analytical mathematician may work from the admitted results of his conic sections, he proceeds to deduce all the conclusions to which he wishes to conduct his readers.

[Praise of Darwin's Writing and Observations]

The essay is full of Mr. Darwin's characteristic excellences. It is a most readable book; full of facts in natural history, old and new, of his collecting and of his observing; and all of these are told in his own perspicuous language, and all thrown into picturesque combinations, and all sparkle with the colours of fancy and the lights of imagination. It assumes, too, the grave proportions of a sustained argument upon a matter of the deepest interest, not to naturalists only, or even to men of science exclusively, but to every one who is interested in the history of man and of the relations of nature around him to the history and plan of creation.

With Mr. Darwin's "argument" we may say in the outset that we shall have much and grave fault to find. But this does not make us the less disposed to admire the singular excellences of his work; and we will seek *in limine* to give our readers a few examples of these. Here, for instance, is a beautiful illustration of the wonderful interdependence of nature—of the golden chain of unsuspected relations which bind together all the mighty web which stretches from end to end of this full and most diversified earth. Who, as he listened to the musical hum of the great hum-ble-bees, or marked their ponderous flight from flower to flower, and watched the unpacking of their trunks for their work of suction, would have supposed that the multiplication or diminution of their race, or the fruitfulness and sterility of the red clover, depend as directly on the vigilance of our cats as do those of our well-guard-ed game-preserves on the watching of our keepers? Yet this Mr. Darwin has discov-ered to be literally the case:—[Quotation from Darwin, p.74.]

Again, how beautiful are the experiments recorded by him concerning that wonder-ful relation of the ants to the aphides, which would almost warrant us in giving to the aphis the name of *Vacca formicaria*:—[Quotation from Darwin, pp 210, 211.]

Or take the following admirable specimen of the union of which we have spoken, of the employment of the observations of others with what he has observed him-self, in that which is almost the most marvelous of facts—the slave-making instinct of certain ants. We say nothing at present of the place assigned to these facts in Mr. Darwin's argument, but are merely referring to the collection, observation, and statement of the facts themselves:—"Slave-making Instinct.—This remarkable instinct was first discovered in the *Formica (Polyerges) rufescens* by Pierre Huber, a better observer even than his celebrated father." [There follows a detailed descrip-tion of slave-making ants and several pages of quotation from Darwin about his observations of ant behavior.]

Now, all this is, we think, really charming writing. We feel as we walk abroad with Mr. Darwin very much as the favoured object of the attention of the dervise must have felt when he had rubbed the ointment around his eye, and had it opened to see all the jewels, and diamonds, and emeralds, and topazes, and rubies, which were sparkling unregarded beneath the earth, hidden as, yet from all eyes save those which the dervise had enlightened. But here we are bound to say our pleasure ter-minates; for, when we turn with Mr. Darwin to his "argument," we are almost immediately at variance with him. It is as an "argument" that the essay is put for-ward; as an argument we will test it.

[Summary of Darwin's Argument]

We can perhaps best convey to our readers a clear view of Mr. Darwin's chain of reasoning, and of our objections to it, if we set before them, first, the conclusion to which he seeks to bring them; next, the leading propositions which he must estab-lish in order to make good his final inference; and then the mode by which he endeavours to support his propositions.

The conclusion, then, to which Mr. Darwin would bring us is, that all the various forms of vegetable and animal life with which the globe is now peopled, or of which we find the remains preserved in a fossil state in the great Earth-Museum around us, which the science of geology unlocks for our instruction, have come down by natural succession of descent from father to son,—"animals from at most four or five progenitors, and plants from an equal or less number" (p. 484), as Mr. Darwin at first somewhat diffidently suggests; or rather, as, growing bolder when he has once pronounced his theory, he goes on to suggest to us, from one single head:—[more quotation from p. 484]

This is the theory which really pervades the whole volume. Man, beast, creeping thing, and plant of the earth, are all the lineal and direct descendants of some one individual *ens*, whose various progeny have been simply modified by the action of natural and ascertainable conditions into the multiform aspect of life which we see around us. This is undoubtedly at first sight a somewhat startling conclusion to arrive at. To find that mosses, grasses, turnips, oaks, worms, and flies, mites and elephants, infusoria and whales, tadpoles of to-day and venerable saurians, truffles and men, are all equally the lineal descendants of the same aboriginal common ancestor, perhaps of the nucleated cell of some primaeval fungus, which alone possessed the distinguishing honour of being the "one primordial form into which life was first breathed by the Creator"—this, to say the least of it, is no common discovery—no very expected conclusion. But we are too loyal pupils of inductive philosophy to start back from any conclusion by reason of its strangeness. Newton's patient philosophy taught him to find in the falling apple the law which governs the silent movements of the stars in their courses; and if Mr. Darwin can with the same correctness of reasoning demonstrate to us our fungular descent, we shall dismiss our pride, and avow, with the characteristic humility of philosophy, our unsuspected cousinship with the mushrooms,— "Claim kindred there, and have our claim allowed,"—only we shall ask leave to scrutinise carefully every step of the argument which has such an ending, and demur if at any point of it we are invited to substitute unlimited hypothesis for patient observation, or the spasmodic fluttering flight of fancy for the severe conclusions to which logical accuracy of reasoning has led the way.

Now, the main propositions by which Mr. Darwin's conclusion is attained are these:—

1. That observed and admitted variations spring up in the course of descents from a common progenitor.

2. That many of these variations tend to an improvement upon the parent stock.

3. That, by a continued selection of these improved specimens as the progenitors of future stock, its powers may be unlimitedly increased.

4. And, lastly, that there is in nature a power continually and universally working out this selection, and so fixing and augmenting these improvements.

Mr. Darwin's whole theory rests upon the truth of these propositions, and crumbles utterly away if only one of them fail him. These therefore we must closely scrutinise. We will begin with the last in our series, both because we think it the newest and the most ingenious part of Mr. Darwin's whole argument, and also because, whilst we absolutely deny the mode in which he seeks to apply the, existence of the power to help him in his argument, yet we think that he throws great and very interesting light upon the fact that such a self-acting power does actively and continuously work in all creation around us.

Mr. Darwin finds then the disseminating and improving power, which he needs to account for the development of new forms in nature, in the principle of "Natural Selection," which is evolved in the strife for room to live and flourish which is evermore maintained between themselves by all living things. One of the most interesting parts of Mr. Darwin's volume is that in which he establishes this law of natural selection; we say establishes, because—repeating that we differ from him totally in the limits which he would assign to its action—we have no doubt of the existence or of the importance of the law itself. Mr. Darwin illustrates it thus:—

"There is no exception to the rule that every organic being naturally increases at so high a rate, that, if not destroyed, the earth would soon be covered by the offspring of a single pair. Linnaeus has calculated that if an annual plant produced only two seeds—and there is no plant so unproductive as this—and their seedlings next year produced two, and so on, then in twenty years there would be a million plants. The elephant is reckoned the slowest breeder of all known animals, and I have taken some pains to estimate its probable minimum rate of natural increase. It will be under the mark to assume that it breeds when thirty years old, and goes on breeding till ninety years old, bringing forth three pair of young in this interval ; if this be so, at the end of the fifth century there would be alive fifteen million elephants, descended from the first pair."—p. 64.

Leaving theoretical calculations, Mr. Darwin proceeds to facts to establish this rapid increase:—[Examples from Darwin follow, pp. 64–68.]

Now all this is excellent. The facts are all gathered from a true observation of nature, and from a patiently obtained comprehension of their undoubted and unquestionable relative significance. That such a struggle for life then actually exists, and that it tends continually to lead the strong to exterminate the weak, we readily admit; and in this law we see a merciful provision against the deterioration, in a world apt to deteriorate, of the works of the Creator's hands. Thus it is that the bloody strifes of the males of all wild animals tend to maintain the vigour and full development of their race; because, through this machinery of appetite and passion, the most vigorous individuals become the progenitors of the next generation of the tribe. And this law, which thus maintains through the struggle of individuals the high type of the family, tends continually, through a similar struggle of species, to lead the stronger species to supplant the weaker.

This indeed is no new observation: Lucretius knew and eloquently expatiated on its truth:—[Latin quotation]. And this, which is true in animal, is no less true in vegetable life. Hardier or more prolific plants, or plants better suited to the soil or conditions of climate, continually tend to supplant others less hardy, less prolific, or less suited to the conditions of vegetable life in those special districts. Thus far, then, the action of such a law as this is clear and indisputable.

But before we can go a step further, and argue from its operation in favour of a perpetual improvement in natural types, we must be shown first that this law of competition has in nature to deal with such favourable variations in the individuals of any species, as truly to exalt those individuals above the highest type of perfection to which their least imperfect predecessors attained—above, that is to say, the normal level of the species;—that such individual improvement is, in truth, a rising above the highest level of any former tide, and not merely the return in its appointed season of the feebler neap to the fuller spring-tide;—and then, next, we must be shown that there is actively at work in nature, co-ordinate with the law of competition and with the existence of "such favourable variations, a power of accumulating such favourable variation through successive descents." Failing the establishment of either of these last two propositions, Mr. Darwin's whole theory falls to pieces. He has accordingly laboured with all his strength to establish these, and into that attempt we must now follow him.

[Argument against Domestic Variations Leading to Speciation]

Mr. Darwin begins by endeavouring to prove that such variations are produced under the selecting power of man amongst domestic animals. Now here we demur *in limine*. Mr. Darwin himself allows that there is a plastic habit amongst domesticated animals which is not found amongst them when in a state of nature. "Under domestication, it may be truly said that the whole organization becomes in some degree plastic"—(p. 80). If so, it is not fair to argue, from the variations of the plastic nature, as to what he himself admits is the far more rigid nature of the undomesticated animal. But we are ready to give Mr. Darwin this point, and to join issue with him on the variations which he is able to adduce, as having been produced under circumstances the most favourable to change. He takes for this purpose the domestic pigeon, the most favourable specimen no doubt, for many reasons, which he could select, as being a race eminently subject to variation, the variations of which have been most carefully observed by breeders, and which, having been for some 4,000 years domesticated, affords the longest possible period for the accumulation of variations. But with all this in his favour, what is he able to show? He writes a delightful chapter upon pigeons. Runts and fantails, short-faced tumblers and long-faced tumblers, long-beaked capriers and pouters, black barbs, jacobins, and turbits, coo and tumble, inflate their oesophagi, and pout and spread out their tail before us. We learn that "pigeons have been watched and tended with the utmost care, and loved by many people." They have been domesticated for thousands of years in several quarters of the world. The earliest known record of pigeons is in the fifth Egyptian dynasty, about 3,000 B.C., though "pigeons are given in a bill of fare" (what an autograph would be that of the chef-de-cuisine of the day!) "in the previous dynasty" (pp. 27, 28): and so we follow pigeons on down to the days of "that most skilful breeder Sir John Sebright," who used to say, with respect to pigeons, that "he would produce any given feather in three years, but it would take him six years to produce beak and head"—(p. 31).

Now all this is very pleasant writing, especially for pigeon-fanciers; but what step do we really gain in it at all towards establishing the alleged fact that variations are but species in the act of formation, or in establishing Mr. Darwin's position that a well-marked variety may be called an incipient species? We affirm positively that no single *fact* tending even in that direction is brought forward. On the contrary, every one points distinctly towards the opposite conclusion; for with all the change wrought in appearance, with all the apparent variation in manners, there is not the faintest beginning of any such change in what that great comparative anatomist, Professor Owen, calls "the characteristics of the skeleton or other parts of the frame upon which specific differences are founded." There is no tendency to that great law of sterility which, in spite of Mr. Darwin, we affirm ever to mark the hybrid; for every variety of pigeon, and the descendants of every such mixture, breed as freely, and with as great fertility, as the original pair; nor is there the very first appearance of that power of accumulating variations until they grow into specific differences, which is essential to the argument for the transmutation of species; for, as Mr. Darwin allows, sudden returns in colour, and other most altered appearances, to the parent stock continually attest the tendency of variations not to become fixed, but to vanish, and manifest the perpetual presence of a principle which leads not to the accumulation of minute variations into well-marked species, but to a return from the abnormal to the original type. So clear is this, that it is well known that any relaxation in the breeder's care effaces all the established points of difference, and the fancy-pigeon reverts again to the character of its simplest ancestor.

The same relapse may moreover be traced in still wider instances. There are many testimonies to the fact that domesticated animals, removed from the care and tending of man, lose rapidly the peculiar variations which domestication had introduced amongst them, and relapse into their old untamed condition. [Examples in French and from other scientists] Now, in all these instances we have the result of the power of selection exercised on the most favourable species for a very long period of time, in a race of that peculiarly plastic habit which is the result of long domestication; and that result is, to prove that there has been no commencement of any such mutation as could, if it was infinitely prolonged, become really a specific change. [Examples of the variety of dogs, all of which are the same species.]

Not let our readers forget over how large a lapse of time our opportunities of observation extend. From the early Egyptian habit of embalming, we know that for 4,000 years at least the species of our own domestic animals, the cat, the dog, and others, has remained absolutely unaltered.

Yet it is in the face of such facts as these that Mr. Darwin ventures, first, to declare that "new races of animals and plants are produced under domestication by man's methodical and unconscious power of selection, for his own use and pleasure," and then to draw from the changes introduced amongst domesticated animals this caution for naturalists: "May they not learn a lesson of caution when they deride the idea of species in a state of nature being lineal descendants of other species?" (p. 29).

[Argument against Variations Leading to Speciation in Nature]

Nor must we pass over unnoticed the transference of the argument from the domesticated to the untamed animals. Assuming that man as the selector can do much in a limited time, Mr. Darwin argues that Nature, a more powerful, a more continuous power, working over vastly extended ranges of time, can do more. But why should Nature, so uniform and persistent in all her operations, tend in this instance to change? Why should she become a selector of varieties? Because, most ingeniously argues Mr. Darwin, in the struggle for life, *if* any variety favourable to the individual were developed, that individual would have a better chance in the battle of life, would assert more proudly his own place, and, handing on his peculiarity to his descendants, would become the progenitor of an improved race; and so a variety would have grown into a species.

We think it difficult to find a theory fuller of assumptions; and of assumptions not grounded upon alleged facts in nature, but which are absolutely opposed to all the facts we have been able to observe.

1. We have already shown that the variations of which we have proof under domestication have never, under the longest and most continued system of selections we have known, laid the first foundation of a specific difference, but have always tended to relapse, and not to accumulated and fixed persistence.

But, 2ndly, all these variations have the essential characteristics of *monstrosity* about them; and *not one* of them has the character which Mr. Darwin repeatedly

Charles Darwin, the Copley Medal, and the Rise of Naturalism

reminds us is the *only one* which nature can select, viz. of being an advantage to the selected individual in the battle of life, "i.e. an improvement upon the normal type by raising some individual of the species not to the highest possible excellence within the species, but to some excellence above it. So far from this, every variation introduced by man is for man's advantage, not for the advantage of the animal. Correlation is so certainly the law of all animal existence that man can only develop one part by the sacrifice of another. The bull-dog gains in strength and loses in swiftness; the grayhound gains in swiftness but loses in strength. Even the English race-horse loses much which would enable it in the battle of life to compete with its rougher ancestor. So too with our prize-cattle. Their greater tendency to an earlier accumulation of meat and fat is counterbalanced, as is well known, by loss of robust health, fertility, and of power of yielding milk, in proportion to their special development in the direction which man's use of them as food requires. There is not a shadow of ground for saying that man's variations ever improve the typical character of the animal as an animal; they do but by some monstrous development make it more useful to himself; and hence it is that Nature, according to her universal law with monstrosities, is ever tending to obliterate the deviation and to return to the type.

The applied argument then, from variation under domestication, fails utterly. But further, what does observation say as to the occurrence of a single instance of such favourable variation? Men have now for thousands of years been conversant as hunters and other rough naturalists with animals of every class. Has any one such instance ever been discovered? We fearlessly assert not one. [Deletion.] Mr. Darwin himself allows that he finds none; and accounts for their absence in existing fauna only by the suggestion, that, in the competition between the less improved parent-form and the improved successor, the parent will have yielded, in the strife in order to make room for the successor; and so "both the parent and all the transitional varieties will generally have been exterminated by the very process of formation and perfection of the new form" (p. 172),—a most unsatisfactory answer as it seems to us; for why—since if this is Nature's law these innumerable changes must be daily occurring—should there never be any one produceable proof of their existence?

Here then again, when subjected to the stern Baconian law of the observation of facts, the theory breaks down utterly; for no natural variations from the specific type favourable to the individual from which nature is to select can anywhere be found.

But once more. If these transmutations were actually occurring, must there not, in some part of the great economy of nature round us, be somewhere at least some instance to be quoted of the accomplishment of the change? With many of the lower forms of animals, life is so short and generations so rapid in their succession that it would be all but impossible, if such changes were happening, that there should be no proof of their occurrence; yet never have the longing observations of Mr. Darwin and the transmutationists found one such instance to establish their theory, and this although the shades between one class and another are often most lightly marked. For there are creatures which occupy a doubtful post between the animal and the vegetable kingdoms—half-notes in the great scale of nature's harmony. Is it credible that all favourable varieties of turnips are tending to become men, and yet that the closest microscopic observation has never detected the faintest tendency in the highest of the Algae to improve into the very lowest Zoophyte?

Again, we have not only the existing tribes of animals out of which to cull, if it were possible, the instances which the transmutationists require to make their theory defensible consistently with the simplest laws of inductive science, but we have in the earth beneath us a vast museum of the forms which have preceded us. Over so vast a period of time does Mr. Darwin extend this collection that he finds reasons for believing that "it is not improbable that a longer period than 300,000,000 years has elapsed since the latter part of the secondary (geological) period alone" (p. 287). Here then surely at last we must find the missing links of that vast chain of innumerable and separately imperceptible variations, which has convinced the inquirer into Nature's undoubted facts of the truth of the transmutation theory. But no such thing. The links are wholly wanting, and the multiplicity of these facts and their absolute rebellion against Mr. Darwin's theory is perhaps his chief difficulty. [Quotes from Darwin.]

[Criticism based on Geological Record]

This "Imperfection of the Geological Record," and the "Geological Succession," are the subjects of two laboured and ingenious chapters, in which he tries, as we think utterly in vain, to break down the unanswerable refutation which is given to his theory by the testimony of the rocks. He treats the subject thus:—1. He affirms that only a small portion of the globe has been explored with care. 2. He extends at will to new and hitherto unsuggested myriads of years the times which have elapsed between successive formations in order to account for the utter absence of everything like a succession of ascertainable variations in the successive inhabitants of the earth. How he deals in these suggestions with time, filling in or striking out a few millions of years at pleasure, the following comprehensive sentence may show:—

"At this rate, on the above data, the denudation of the Weald must have required 306,662,400 years, or say three hundred million years. But perhaps it would be safer to allow two or three inches per century, and this would reduce the number of years to 150 or 100 million years."—p. 287.

As these calculations concerning the general duration of formations, and specially concerning the Weald, are highly characteristic of the whole "argument," it may be worth while to submit them to a somewhat closer examination.

Mr. Darwin then argues (pp. 285, 286) that "faults" proclaim the vastness of these durations. To establish this, he supposes that the result of a great fracture was the severing of strata once continuous, so as to throw them relatively a thousand feet apart from their original position, and thus form a cliff which stood up vertically on one side of that dislocation; and so he imagines that countless ages must have elapsed, according to the present waste of land, to account for the wearing down of these outlines, so as to have left (as is often the case) no trace of the great dislocation upon the present surface of the land. But, with hardly an exception, every sound geologist would repudiate as a "petitio principii" this whole method of reasoning; for though a few geologists would explain these great dislocations on the hypothesis of intermittent successive movements severally of small amount, yet in the judgment of far the larger number, and the more judicious of those who have made geology their study, they were undoubtedly the result of sudden movements, produced by internal efforts of central heat and gas to escape, and were infinitely more intense and spasmodic (catastrophic if you will) than any of those similar causes which, in a minor way, now produce our earthquakes and oscillations of the surface to the extent of a few feet only. Hence these great breaks and fractures were of such a nature as to render it impossible that any cliff should, at the period of their formation, have stood up on one side of the fracture. The very violence of the movement, accompanied as it must have been by the translation of vast masses of water sweeping away the rubbish, may, on the instant, have almost entirely smoothed down the ruptured fragments; the more so, as most of these great dislocations are believed to have taken place under the sea. The flattening down of all superficial appearances was therefore most probably the direct result of the catastrophe, and the countless ages of Darwin were, in all probability, at the longest, nothing more than a few months or years of our time.

The whole argument as to the Wealden denudation (p. 287) appears to us a similar exaggeration. Granting that rocky coasts are very slowly worn away by the present sea, the application of this view to the north and south coasts of the valley of the Weald, i.e. to the escarpments of the North and South Downs, is entirely untenable. For what shadow of proof is there that these chalk escarpments have been worn down inch by inch by the erosion of the waves of a former sea? It may be said to have been demonstrated by that great practical observer and philosophical geologist Sir R. Murchison, that, inasmuch as there is no trace of rounded water-worn pebbles nor shingles in any portion of the Weald (though there were plenty on the slopes without), the sea never could have so acted along these escarpments as on a shore, and hence the whole of the basis of the reasoning, about the three hundred million of years for the denudation of the cretaceous and subjacent deposits, is itself washed away at once.

But not only do the facts to which Mr. Darwin trusts to establish his vast lapses of years, which, he says, "impress his mind almost in the same manner as does the vain endeavour to grapple with the idea of Eternity" (p. 285), not only do these give him the same power of supposing the progress of changes, of which we have found neither the commencement, nor the progress, nor the record, as ancient geographers allowed themselves, when they speculated upon the forms of men whose heads grew beneath their shoulders in the unreached recesses of Africa,— but when, passing from these unlimited terms for change to work in, he proceeds to deal with the absence of all record of the changes themselves, the plainest geological facts again disprove his assumptions. For here lie assumptions that there are everywhere vast gaps (p. 302) between successive formations, which might, if they were filled up, furnish instances of all the many gradations required by his theory, and also that the past condition of the earth made the preservation of such specimens improbable. To prove the existence of these wide gaps, Mr. Darwin quotes (p. 289) Sir R. Murchison's great work on "Russia"; but he appears to us to quote it incorrectly, for we understand it to say that there is abundant evidence that in that drift-covered region there are many evidences of the transition from the Devonian into the Carboniferous era in Palaeozoic life, and also from the old Aralo-Caspian, or brackish water condition of tertiary times into present oceanic life; and that if all the rocks of Russia could be uncovered and the drift removed, we might discover many more of these transitions. In fact, although the geological record is often broken, we already know of many unbroken and perfect transitions between the Cambrian and Silurian, between the Silurian and Devonian, between the Devonian and Carboniferous, if not between the latter and the Permian.

Again, there is an absolute unbroken physical connection in Germany between the Permian and the Trias, and yet an entire separation of animals, and so on in Secondary and Tertiary deposits.

Now, if the field-geologist can show clear proofs of continuous deposit, and yet many distinct plants and animals in the succeeding formations, what becomes of that immense lapse of ages which should transform the Palaeozoic Permian type into the entirely distinct Secondary or Triassic form? All such links are absolutely wanting even in these tracts, and in many others, where the conformable and gradual transition between formations proves that there is between them no break, and where everything indicates quiet physical transition, and which yet contain utterly different remains. How then can we account for such distinct forms of life in the quietly succeeding formations except by distinct creations?

Mr. Darwin is compelled to admit that he finds no records in the crust of the earth to verify his assumption:—[quotes Darwin on inadequate geological record, p. 308.] As to the suggestion that the absence of organic remains is no proof of the non-existence of the unrepresented classes, we would rather speak in the weighty words of Professor Owen than employ our own:—[quotes from Owen's *Classification of Mammalia*, pp. 58–60.]

Charles Darwin, the Copley Medal, and the Rise of Naturalism

Mr. Darwin's own pages bear witness to the same conclusion. The rare land shell found by Sir C. Lyell and Dr. Dawson in North America affords a conclusive proof that in the carboniferous period such animals were most rare, and only the earliest of that sort created. For the carboniferous strata of North America, stretching over tracts as large as the British Isles, and containing innumerable plants and other terrestrial things, must have been very equally depressed and elevated, since the very flowers and fruits of the plants of the period have been preserved; and if terrestrial animals abounded, why do we not see more of their remains than this miserable little dendropupa about a quarter of an inch long?

It would be wearisome to prolong these proofs; but if to any man they seem insufficient, let him read carefully the conclusion of Sir Roderick Murchison's masterly work upon "Siluria." We venture to aver that the conviction must be forced upon him that the geological record is absolutely inconsistent with the truth of Mr. Darwin's theory; and yet by Mr. Darwin's own confession this conclusion is fatal to his whole argument:—

"If my theory be true, it is indisputable that, before the lowest Silurian stratum was deposited, long periods elapsed, as long as, or probably far longer than, the whole interval from the Silurian age to the present day; and that during these vast yet quite unknown periods of time, the world swarmed with living creatures."—p. 307.

Now it is proved to demonstration by Sir Roderick Murchison, and admitted by all geologists, that we possess these earlier formations, stretching over vast extents, perfectly unaltered, and exhibiting no signs of life. Here we have, as nearly as it is possible in the nature of things, to have, the absolute proof of a negative. If these forms of life had existed they must have been found. Even Mr. Darwin shrinks from the deadly gripe of this argument. "The case," he says (p. 308) "at present must remain inexplicable, and may be truly urged as a valid argument against the views here entertained." More than once indeed does he make this admission. One passage we have quoted already from p. 280 of his work. With equal candour he says further on:—[quotes Darwin, pp. 302, 463.]

[Argument based on sterility of hybrids]

But though this objection is that which is rated highest by himself, there is another which appears to us in some respects stronger still, and to which we deem Mr. Darwin's answers equally insufficient,—we mean the law of sterility affixed to hybridism. If it were possible to proclaim more distinctly by one provision than another that the difference between various species was a law of creation, and not, as the transmutationists maintain, an ever-varying accident, it would surely be by the interposing such a bar to change as that which now exists in the universal fruitlessness which is the result of all known mixtures of animals specifically distinct.

Mr. Darwin labours hard here, but his utmost success is to reveal a very few instances from the vegetable world, with its shadowy image of the procreative animal system, as exceptions to the universal rule. As to animals, he is compelled by the plainness of the testimony against him to admit that he "doubts whether any case of a perfectly fertile hybrid animal can be considered as thoroughly well authenticated" (p. 252); and his best attempts to get rid of this evidence are such suggestions as that "the common and the true ring-necked pheasant intercross" (p. 253), though every breeder of game could tell him that, so far from there being the slightest ground for considering these as distinct species, all experience shows that the ring-neck almost uniformly appears where the common pheasant's eggs are hatched under the domestic hen. How then does Mr. Darwin dispose of this apparently impassable barrier of nature against the transmutation-theory? He urges that it depends not upon any great law of life, but mainly, first, on the early death of the embryo, or, secondly, upon "the common imperfection of the reproductive system" in the male offspring. How he considers this to be any answer to the difficulty it is beyond our power to conceive. We can hardly imagine any clearer way of stating the mode in which an universal law, if it existed, must act, than that in which he describes it, to disprove its existence. But, besides this, other and insuperable difficulties beset this whole speculation. To one of these Mr. Darwin alludes (pp. 192, 193), and dismisses it with a most suspicious brevity. "The electric organs of fishes," he says, "offer another case of special difficulty," and he places as "a parallel case of difficulty the presence of luminous organs in a few insects belonging to different families and orders.

We see no possible solution on the Darwinian theory for the presence at once so marked and so exceptional of these organs. And how are they dealt with? Surely in a mode most unsatisfactory by one promulging a new theory of creation; for scarcely admitting that their presence is little else than destructive of his theory, Mr. Darwin simply remarks "that we are too ignorant to argue that no transition of any kind is possible," a solution which could of course equally make the scheme it is intended to serve compatible with any other contradiction.

[Argument regarding Special Case of Poisonous Animals]

It is the more important to notice this, because there is another large class of cases in which the same difficulty is present, and as to which Mr. Darwin suggests no solution. We allude to those animals which, like many snakes, possess special organs for secreting venom and for discharging it at their own proper volition. The whole set of glands, ducts, and other vessels employed for this purpose are, as any instructed comparative anatomist would tell him, so entirely separate from the ordinary laws of animal life and peculiar to themselves, that the derivation of these by any natural modification from progenitors which did not possess them would be a marvellous contradiction of all laws of descent with which we are familiar. And this special and unnoticed difficulty leads us on to another of still wider extent. Most of our readers know that the stomachs and whole digestive system of the carnivori are constructed upon a wholly different type from those of the graminivorous animals. Yet whence this difference, if these diverse constructions can claim a common origin? Can any permutationist pretend that experience gives us any reason for believing that any change of food, however unnatural or forced, ever has changed or ever could change the one type into the other? Yet that diversity pervades the whole being of the separated classes. It does not affect only their outward forms, as to which the merest accidents of colour or of hair may veil real resemblance under seeming difference, but it pervades the nervous system, the organs of reproduction, the stomach, the alimentary canal; nay, in every blood-corpuscle which circulates through their arteries and veins it is universally present and perpetually active.

Where, then, in the most allied forms, was the earliest commencement of diversity? or what advantage of life could alter the shape of the corpuscles into which the blood can be evaporated?

We come then to these conclusions. All the facts presented to us in the natural world tend to show that none of the variations produced in the fixed forms of animal life, when seen in its most plastic condition under domestication, give any promise of a true transmutation of species; first, from the difficulty of accumulating and fixing variations within the same species; secondly, from the fact that these variations, though most serviceable for man, have no tendency to improve the individual beyond the standard of his own specific type, and so to afford matter, even if they were infinitely produced, for the supposed power of natural selection on which to work; whilst all variations from the mixture of species are barred by the inexorable law of hybrid sterility. Further, the embalmed records of 3000 years show that there has been no beginning of transmutation in the species of our most familiar domesticated animals; and beyond this, that in the countless tribes of animal life around us, down to its lowest and most variable species, no one has ever discovered a single instance of such transmutation being now in prospect; no new organ has ever been known to be developed—no new natural instinct to be formed—whilst, finally, in the vast museum of departed animal life which the strata of the earth imbed for our examination, whilst they contain far too complete a representation of the past to be set aside as a mere imperfect record, yet afford no one instance of any such change as having ever been in progress, or give us anywhere the missing links of the assumed chain, or the remains which would enable now existing variations, by gradual approximations, to shade off into unity.

[Argument against "Hypothesis" as opposed to "Induction"]

On what then is the new theory based? We say it with unfeigned regret, in dealing with such a man as Mr. Darwin, on the merest hypothesis, supported by the most unbounded assumptions. These are strong words, but we will give a few instances to prove their truth. [quotes Darwin, pp. 191, 204, 181, 192–3, 186, 187, 189 showing his speculations.]

Another of these assumptions is not a little remarkable. It suits the argument to deduce all our known varieties of pigeon from the rock-pigeon (the Columba livia), and this parentage is traced out, though not, we think, to demonstration, yet with great ingenuity and patience. But another branch of the argument would be greatly strengthened by establishing the descent of our various breeds of dogs with their perfect power of fertile interbreeding from different natural species. And accordingly, though every fact as to the canine race is parallel to the facts which have been used before to establish the common parentage of the pigeons in Columba livia, all these are thrown over in a moment, and Mr. Darwin, first assuming, without the shadow of proof, that our domestic breeds are descended from different species, proceeds calmly to argue from this, as though it were a demonstrated certainty.

"It *seems to me unlikely* in the case of the dog-genus, which is distributed in a wild state throughout the world, that since man first appeared one species alone should have been domesticated."—p. 18.

"In some cases I do not doubt that the intercrossing of species aboriginally distinct has played an important part in the origin of our domestic productions."—p. 43.

What new words are these for a loyal disciple of the true Baconian philosophy?—"I can conceive"—"It is not incredible"—"I do not doubt"—"It is conceivable."

"For myself, *I venture confidently* to look back thousands on thousands of generations, and I see an animal striped like a zebra, but perhaps otherwise very differently constructed, the common parent of our domestic horse, whether or not it be descended from one or more wild stocks of the ass, heminus, quagga, or zebra."—p. 167.

In the name of all true philosophy we protest equally against such a mode of dealing with nature, as utterly dishonourable to all natural science, as reducing it from its present lofty level as one of the noblest trainers of man's intellect and instructors of his mind, to being a mere idle play of the fancy, without the basis of fact or the discipline of observation. In the "Arabian Nights" we are not offended as at an impossibility when Amina sprinkles her husband with water and transforms him into a dog, but we cannot open the august doors of the venerable temple of scientific truth to the genii and magicians of romance. We plead guilty to Mr. Darwin's imputation that the chief cause of our natural unwillingness to admit that one species has given birth to other and distinct species is that we are always slow in admitting any great change of which we do not see the intermediate steps."—p. 481.

In this tardiness to admit great changes suggested by the imagination, but the steps of which we cannot see, is the true spirit of philosophy.

"Analysis," says Professor Sedgwick, "consists in making experiments and observations, and in drawing general conclusions from them by induction, and admitting of no objections against the conclusions but such as are taken from experiments or other certain truths; for hypotheses are not to be regarded in experimental philosophy."

The other solvent which Mr. Darwin most freely and, we think, unphilosophically employs to get rid of difficulties, is his use of time. This he shortens or prolongs at will by the mere wave of his magician's rod. Thus the duration of whole epochs, during which certain forms of animal life prevailed, is gathered up into a point, whilst an unlimited expanse of years, impressing his mind with a sense of eternity, is suddenly interposed between that and the next series, though geology proclaims the transition to have been one of gentle and, it may be, swift accomplishment. All this too is made the more startling because it is used to meet the objections drawn from facts. "We see none of your works," says the observer of nature; "we see no beginnings of the portentous change; we see plainly beings of another order in creation, but we find amongst them no tendencies to these altered organisms." True says the great magician, with a calmness no difficulty derived from the obstinacy of facts can disturb; "true, but remember the effect of time. Throw in a few hundreds of millions of years more or less, and why should not all these changes be possible, and, if possible, why may I not assume them to be real?"

Together with this large licence of assumption we notice in this book several instances of receiving as facts whatever seems to bear out the theory upon the slightest evidence, and rejecting summarily others, merely because they are fatal to it. We grieve to charge upon Mr. Darwin this freedom in handling facts, but truth extorts it from us. That the loose statements and unfounded speculations of this book should come from the author of the monograms on Cirripedes, and the writer, in the natural history of the Voyage of the "Beagle," of the paper on the Coral Reefs, is indeed a sad warning how far the love of a theory may seduce even a first-rate naturalist from the very articles of his creed.

[Argument in Favor of Independent Creation]

This treatment of facts is followed up by another favourite line of argument, namely, that by this hypothesis difficulties otherwise inextricable are solved. Such passages abound. Take a few, selected almost at random, to illustrate what we mean:— [quotes Darwin directly contrasting explanation of natural selection with independent creation, pp. 436, 477–8, 398–9.]

Now what can be more simply reconcilable with that theory than Mr. Darwin's own account of the mode in which the migration of animal life from one distant region to another is continually accomplished?

Take another of these suggestions:—

"It is inexplicable, on the theory of creation, why a part developed in a very unusual manner in any one species of a genus, and therefore, as we may naturally infer, of great importance to the species, should be eminently liable to variation."—p. 474.

Why "inexplicable"? Such a liability to variation might most naturally be expected in the part "unusually developed," because such unusual development is of the nature of a monstrosity, and monsters are always tending to relapse into likeness to the normal type. Yet this argument is one on which he mainly relies to establish his theory, for he sums all up in this triumphant inference:—

"I cannot believe that a false theory would explain, as it seems to me that the theory of natural selection does explain, the several large classes of facts above specified."—p, 480.

Now, as to all this, we deny, first, that many of these difficulties are "inexplicable on any other supposition." Of the greatest of them (128, 194) we shall have to speak before we conclude. We will here touch only on one of those which are continually reappearing in Mr. Darwin's pages, in order to illustrate his mode of dealing with them. He finds, then, one of these "inexplicable difficulties" in the fact, that the young of the blackbird, instead of resembling the adult in the colour of its plumage, is like the young of many other birds spotted, and triumphantly declaring that—

"No one will suppose that the stripes on the whelp of a lion, or the spots on the young blackbird, are of any use to these animals, or are related to the conditions to which they are exposed"—pp. 439–40.

He draws from them one of his strongest arguments for this alleged community of descent. Yet what is more certain to every observant field-naturalist than that this alleged uselessness of colouring is one of the greatest protections to the young bird, imperfect in its flight, perching on every spray, sitting unwarily on every bush through which the rays of sunshine dapple every bough to the colour of its own plumage, and so give it a facility of escape which it would utterly want if it bore the marked and prominent colours, the beauty of which the adult bird needs to recommend him to his mate, and can safely bear with his increased habits of vigilance and power of wing?

But, secondly, as to many of these difficulties, the alleged solving of which is one great proof of the truth of Mr. Darwin's theory, we are compelled to join issue with him on another ground, and deny that he gives us any solution at all. Thus, for instance, Mr. Darwin builds a most ingenious argument on the tendency of the young of the horse, ass, zebra, and quagga, to bear on their shoulder and on their legs certain barred stripes. Up these bars (bars sinister, as we think, as to any true descent of existing animals from their fancied prototype) he mounts through his "thousands and thousands of generations," to the existence of his "common parent, otherwise perhaps very differently constructed, but striped like a zebra."—(p. 67.) "How inexplicable," he exclaims, "on the theory of creation, is the occasional appearance of stripes on the shoulder and legs of several species of the horse genus and in their hybrids!"—(p. 473.) He tells us that to suppose that each species was created with a tendency "like this, is to make the works of God a mere mockery and deception"; and he satisfies himself that all difficulty is gone when he refers the stripes to his hypothetical thousands on thousands of years removed progenitor. But how is his difficulty really affected for why is the striping of one species a less real difficulty than the striping of many?

Another instance of this want of fairness, to which we must call the attention of our readers, because it too often recurs, is contained in the following question:—

"Were all the infinitely numerous kinds of animals and plants created as eggs, or seed, or as full grown? and, in the case of mammals, were they created bearing the false marks of nourishment from the mother's womb?"—p. 483.

The difficulty here glanced at is extreme, but it is one for the solution of which the transmutation-theory gives no clue. It is inherent in the idea of the creation of beings, which are to reproduce their like by natural succession; for, in such a world, place the first beginning where you will, that beginning must contain the apparent history of a *past*, which existed only in the mind of the Creator. If, with Mr. Darwin, to escape the difficulty of supposing the first man at his creation to possess in that framework of his body "false marks of nourishment from his mother's womb," with Mr. Darwin you consider him to have been an improved ape, you only carry the difficulty up from the first man to the first ape; if, with Mr. Darwin, in violation of all observation, you break the barrier between the classes of vegetable and animal life, and suppose every animal to be an "improved" vegetable, you do but carry your difficulty with you into the vegetable world; for, how could there be seeds if there had been no plants to seed them? and if you carry up your thoughts through the vista of the Darwinian eternity up to the primaeval fungus, still the primaeval fungus must have had humus, from which to draw into its venerable vessels the nourishment of its archetypal existence, and that humus must itself be a "false mark" of a preexisting vegetation.

We have dwelt a little upon this, because it is by such seeming solutions of difficulties as that which this passage supplies that the transmutationist endeavours to prop up his utterly rotten fabric of guess and speculation.

[Argument against Improving Instinct]

There are no parts of Mr. Darwin's ingenious book in which he gives the reins more completely to his fancy than where he deals with the improvement of instinct by his principle of natural selection. We need but instance his assumption, without a fact on which to build it, that the marvellous skill of the honey-bee in constructing its cells is thus obtained, and the slave-making habits of the *Formica Polyerges* thus formed. There seems to be no limit here to the exuberance of his fancy, and we cannot but think that we detect one of those hints by which Mr. Darwin indicates the application of his system from the lower animals to man himself, when he dwells so pointedly upon the fact that it is always the black ant which is enslaved by his other coloured and more fortunate brethren. "The slaves are black!" We believe that, if we had Mr. Darwin in the witness-box, and could subject him to a moderate cross-examination, we should find that he believed that the tendency of the lighter-coloured races of mankind to prosecute the negro slave-trade was really a remains, in their more favoured condition, of the "extraordinary and odious instinct" which had possessed them before they had been "improved by natural selection" from *Formica Polyerges* into Homo. This at least is very much the way in which (p. 479) he slips in quite incidentally the true identity of man with the horse, the bat, and the porpoise:—

"The framework of bones being the same in the hand of a man, wing of a bat, fin of a porpoise, and leg of the horse, the same number of vertebrae forming the neck of the giraffe and of the elephant, and innumerable other such facts, at once explain themselves on the theory of descent with slow and slight successive modifications."—p. 479.

Such assumptions as these, we once more repeat, are most dishonourable and injurious to science; and though, out of respect to Mr. Darwin's high character and to the tone of his work, we have felt it right to weigh the "argument" again set by him before us in the simple scales of logical examination, yet we must remind him that the view is not a new one, and that it has already been treated with admirable humour when propounded by another of his name and of his lineage. We do not think that, with all his matchless ingenuity, Mr. Darwin has found any instance which so well illustrates his own theory of the improved descendant under the elevating influences of natural selection exterminating the progenitor whose specialities he has exaggerated as he himself affords us in this work. For if we go back two generations we find the ingenious grandsire of the author of the *On the Origin of Species* speculating on the same subject, and almost in the same manner with his more daring descendant. [Lengthy section reiterating criticisms of Erasmus Darwin's work.]

[Addressing the Religious Issue]

Our readers will not have failed to notice that we have objected to the views with which we have been dealing solely on scientific grounds. We have done so from our fixed conviction that it is thus that the truth or falsehood of such arguments should be tried. We have no sympathy with those who object to any facts or alleged facts in nature, or to any inference logically deduced from them, because they believe them to contradict what it appears to them is taught by Revelation. We think that all such objections savour of a timidity which is really inconsistent with a firm and well-instructed faith:—

"Let us for a moment," profoundly remarks Professor Sedgwick, "suppose that there are some religious difficulties in the conclusions of geology. How, then, are we to solve them? Not by making a world after a pattern of our own—not by shifting and shuffling the solid strata of the earth, and then dealing them out in such a way as to play the game of an ignorant or dishonest hypothesis—not by shutting our eyes to facts, or denying the evidence of our senses—but by patient investigation, carried on in the sincere love of truth, and by learning to reject every consequence not warranted by physical evidence."

He who is as sure as he is of his own existence that the God of Truth is at once the God of Nature and the God of Revelation, cannot believe it to be possible that His voice in either, rightly understood, can differ, or deceive His creatures. To oppose facts in the natural world because they seem to oppose Revelation, or to humour them so as to compel them to speak its voice is, he knows, but another form of the ever-ready feebleminded dishonesty of lying for God, and trying by fraud or falsehood to do the work of the God of truth. It is with another and a nobler spirit that the true believer walks amongst the works of nature. The words graven on the everlasting rocks are the words of God, and they are graven by His hand. No more can they contradict His Word written in His book, than could the words of the old covenant graven by His hand on the stony tables contradict the writings of His hand in the volume of the new dispensation. There may be to man difficulty in reconciling all the utterances of the two voices. But what of that? He has learned already that here he knows only in part, and that the day of reconciling all apparent contradictions between what must agree is nigh at hand. He rests his mind in perfect quietness on this assurance, and rejoices in the gift of light without a misgiving as to what it may discover:—

"A man of deep thought and great practical wisdom," says Sedgwick, "one whose piety and benevolence have for many years been shining before the world, and of whose sincerity no scoffer (of whatever school) will dare to start a doubt, recorded his opinion in the great assembly of the men of science who during the past year were gathered from every corner of the Empire within the walls of this University, 'that Christianity had everything to hope and nothing to fear from the advancement of philosophy.'"

This is as truly the spirit of Christianity as it is that of philosophy. Few things have more deeply injured the cause of religion than the busy fussy energy with which men, narrow and feeble alike in faith and in science, have bustled forth to reconcile all new discoveries in physics with the word of inspiration. For it continually happens that some larger collection of facts, or some wider view of the phenomena of nature, alter the whole philosophic scheme; whilst Revelation has been committed to declare an absolute agreement with what turns out after all to have been a misconception or an error. We cannot, therefore, consent to test the truth of natural science by the Word of Revelation. But this does not make it the less important to point out on scientific grounds scientific errors, when those errors tend to limit God's glory in creation, or to gainsay the revealed relations of that creation to Himself. To both these classes of error, though, we doubt not, quite unintentionally on his part, we think that Mr. Darwin's speculations directly tend.

Mr. Darwin writes as a Christian, and we doubt not that he is one. We do not for a moment believe him to be one of those who retain in some corner of their hearts a secret unbelief which they dare not vent ; and we therefore pray him to consider well the grounds on which we brand his speculations with the charge of such a tendency. First, then, he not obscurely declares that he applies his scheme of the action of the principle of natural selection to MAN himself, as well as to the animals around him.

Now, we must say at once, and openly, that such a notion is absolutely incompatible not only with single expressions in the word of God on that subject of natural science with which it is not immediately concerned, but, which in our judgment is of far more importance, with the whole representation of that moral and spiritual condition of man which is its proper subject-matter. Man's derived supremacy over the earth; man's power of articulate speech; man's gift of reason; man's free-will and responsibility; man's fall and man's redemption; the incarnation of the Eternal Son; the indwelling of the Eternal Spirit,—all are equally and utterly irreconcilable with the degrading notion of the brute origin of him who was created in the image of God, and redeemed by the Eternal Son assuming to himself his nature. Equally inconsistent, too, not with any passing expressions, but with the whole scheme of God's dealings with man as recorded in His word, is Mr. Darwin's daring notion of man's further development into some unknown extent of powers, and shape, and size, through natural selection acting through that long vista of ages which he casts mistily over the earth upon the most favoured individuals of his species. We care not in these pages to push the argument further. We have done enough for our purpose in thus succinctly intimating its course. If any of our readers doubt what must be the result of such speculations carried to their logical and legitimate conclusion, let them turn to the pages of Oken, and see for themselves the end of that path the opening of which is decked out in these pages with the bright hues and seemingly innocent deductions of the transmutation-theory.

Nor can we doubt, secondly, that this view, which thus contradicts the revealed relation of creation to its Creator, is equally inconsistent with the fulness of His glory. It is, in truth, an ingenious theory for diffusing throughout creation the working and so the personality of the Creator. And thus, however unconsciously to him who holds them, such views really tend inevitably to banish from the mind most of the peculiar attributes of the Almighty.

How, asks Mr. Darwin, can we possibly account for the manifest plan, order, and arrangement which pervade creation, except we allow to it this self-developing power through modified descent?

"As Milne-Edwards has well expressed it, Nature is prodigal in variety, but niggard in innovation. Why, on the theory of creation, should this be so? Why should all the parts and organs of many independent beings, each supposed to have been separately created for its proper place in nature, be so commonly linked together by graduated steps? Why should not Nature have taken a leap from structure to structure?"—p. 194.

And again:—

"It is a truly wonderful fact—the wonder of which we are apt to overlook from familiarity—that all animals and plants throughout all time and space should be related to each other in group subordinate to group, in the manner which we everywhere behold, namely, varieties of the same species most closely related together, species of the same genus less closely and unequally related together, forming sections and sub-genera, species of distinct genera much less closely related, and genera related in different degrees, forming sub-families, families, orders, sub-classes, and classes."—pp. 128-9.

How can we account for all this? By the simplest and yet the most comprehensive answer. By declaring the stupendous fact that all creation is the transcript in matter of ideas eternally existing in the mind of the Most High—that order in the utmost perfectness of its relation pervades His works, because it exists as in its centre and highest fountain-head in Him the Lord of all. Here is the true account of the fact which has so utterly misled shallow observers, that Man himself, the Prince and Head of this creation, passes in the earlier stages of his being through phases of existence closely analogous, so far as his earthly tabernacle is concerned, to those in which the lower animals ever remain. At that point of being the development of the protozoa is arrested. Through it the embryo of their chief passes to the perfection of his earthly frame. But the types of those lower forms of being must be found in the animals which never advance beyond them—not in man for whom they are but the foundation for an after-development; whilst he too, Creation's crown and perfection, thus bears witness in his own frame to the law of order which pervades the universe.

In like manner could we answer every other question as to which Mr. Darwin thinks all oracles are dumb unless they speak his speculation. He is, for instance, more than once troubled by what he considers imperfections in Nature's work. "If," he says, "our reason leads us to admire with enthusiasm a multitude of inimitable contrivances in Nature, this same reason tells us that some other contrivances are less perfect."

"Nor ought we to marvel if all the contrivances in nature be not, as far as we can judge, absolutely perfect; and if some of them be abhorrent to our idea of fitness. We need not marvel at the sting of the bee causing the bee's own death; at drones being produced in such vast numbers for one single act, with the great majority slaughtered by their sterile sisters; at the astonishing waste of pollen by our fir-trees; at the instinctive hatred of the queen-bee for her own fertile daughters; at ichneumonidse feeding within the live bodies of caterpillars; and at other such cases. The wonder indeed is, on the theory of natural selection, that more cases of the want of absolute perfection have not been observed."—p. 472.

We think that the real temper of this whole speculation as to nature itself may be read in these few lines. It is a dishonouring view of nature.

That reverence for the work of God's hands with which a true belief in the All-wise Worker fills the believer's heart is at the root of all great physical discovery; it is the basis of philosophy. He who would see the venerable features of Nature must not seek with the rudeness of a licensed roysterer violently to unmask her countenance; but must wait as a learner for her willing unveiling. There was more of the true temper of philosophy in the poetic fiction of the Panic shriek, than in the atheistic speculations of Lucretius. But this temper must beset those who do in effect banish God from nature. And so Mr. Darwin not only finds in it these bungling contrivances which his own greater skill could amend, but he stands aghast before its mightier phenomena. The presence of death and famine seems to him inconceivable on the ordinary idea of creation; and he looks almost aghast at them until reconciled to their presence by his own theory that "a ratio of increase so high as to lead to a struggle for life, and as a consequence to natural selection entailing divergence of character and the extinction of less improved forms, is decidedly followed by the most exalted object which we are capable of conceiving, namely, the production of the higher animals" (p. 490). But we can give him a simpler solution still for the presence of these strange forms of imperfection and suffering amongst the works of God.

We can tell him of the strong shudder which ran through all this world when its head and ruler fell. When he asks concerning the infinite variety of these multiplied works which are set in such an orderly unity, and run up into man as their reasonable head, we can tell him of the exuberance of God's goodness and remind him of the deep philosophy which lies in those simple words—"All thy works praise Thee, O God, and thy saints give thanks unto Thee." For it is one office of redeemed man to collect the inarticulate praises of the material creation, and pay them with conscious homage into the treasury of the supreme Lord. Surely the philosophy which penned the following glorious words is just as much truer to nature as it is to revelation than all these speculations of the transmutationist. Having shown, from a careful osteological examination of his structure, from his geographical distribution, from the differences and agreements of the several specimens of the human family, and from the changes which step by step we can trace wrought by domestication and variation in the lower animals, that man is not and cannot be an improved ape, Professor Owen adds:— [quotes from both Owen and Darwin, contrasting their views on past and future possible changes in species]

It is by putting . . . restraint upon fancy that science is made the true trainer of our intellect:—

"A study of the Newtonian philosophy," says Sedgwick, "as affecting our moral powers and capacities, does not terminate in mere negations. It teaches us to see the finger of God in all things animate and inanimate, and gives us an exalted conception of His attributes, placing before us the clearest proof of their reality; and so prepares, or ought to prepare, the mind for the reception of that higher illumination which brings the rebellious faculties into obedience to the Divine will."—*Studies of the University*, p. 14.

It is by our deep conviction of the truth and importance of this view for the scientific mind of England that we have been led to treat at so much length Mr. Darwin's speculation. The contrast between the sober, patient, philosophical courage of our home philosophy, and the writings of Lamarck and his followers and predecessors, of MM. Demailet, Bory de Saint Vincent, Virey, and Oken, is indeed most wonderful; and it is greatly owing to the noble tone which has been given by those great men whose words we have quoted to the school of British science. That Mr. Darwin should have wandered from this broad highway of nature's works into the jungle of fanciful assumption is no small evil. We trust that he is mistaken in believing that he may count Sir C. Lyell as one of his converts. We know indeed the strength of the temptations which he can bring to bear upon his geological brother. The Lyellian hypothesis, itself not free from some of Mr. Darwin's faults, stands eminently in need for its own support of some such new scheme of physical life as that propounded here. Yet no man has been more distinct and more logical in the denial of the transmutation of species than Sir C. Lyell, and that not in the infancy of his scientific life, but in its full vigour and maturity. [arguments from Lyell] . . . He urges:—

1. That there is a capacity in all species to accommodate themselves to a certain extent to a change of external circumstances.

4. The entire variation from the original type . . . may usually be effected in a brief period of time, after which no further deviation can be obtained.

5. The intermixing of distinct species is guarded against by the sterility of the mule offspring.

6. It appears that species have a real existence in nature, and that each was endowed at the time of its creation with the attributes and organization by which it is now distinguished.

We trust that Sir C. Lyell abides still by these truly philosophical principles; and that with his help and with that of his brethren this flimsy speculation may be as completely put down as was what in spite of all denials we must venture to call its twin though less-instructed brother, the "Vestiges of Creation." In so doing they will assuredly provide for the strength and continually growing progress of British science.

Indeed, not only do all laws for the study of nature vanish when the great principle of order pervading and regulating all her processes is given up, but all that imparts the deepest interest in the investigation of her wonders will have departed too. Under such influences man soon goes back to the marvelling stare of childhood at the centaurs and hippogriffs of fancy, or if he is of a philosophic turn, he comes like Oken to write a scheme of creation under "a sort of inspiration"; but it is the frenzied inspiration of the inhaler of mephitic gas. The whole world of nature is laid for such a man under a fantastic law of glamour, and he becomes capable of believing anything: to him it is just as probable that Dr. Livingstone will find the next tribe of negroes with their heads growing under their arms as fixed on the summit of the cervical vertebrae; and he is able, with a continually growing neglect of all the facts around him, And with equal confidence and equal delusion, to look back to any past and to look on to any future.

JOHN LUBBOCK, "TACT"

[The following version of has been abridged and annotated from *Tact*, Jersey City, NJ: Wells Publishing, 1933, pp. 11–25. Editor's notes appear in brackets [ ].]

For success in life tact is more important than talent, but it is not easily acquired by those to whom it does not come naturally. Still something can be done by considering what others would probably wish. Never lose a chance of giving pleasure. Be courteous to all. "Civility," said Lady Montague, "costs nothing and buys everything." It buys much, indeed, which no money will purchase. Try then to win every one you meet. "Win their hearts," said Burleigh to Queen Elizabeth, "and you have all men's hearts and purses."

Tact often succeeds where force fails. Lilly quotes the old fable of the Sun and the Wind: "It is pretily noted of a contention betweene the Winde and the Sunne, who should have the victorye. A Gentleman walking abroad, the Winde thought to blowe off his cloake, which with great blastes and blusterings striving to unloose it, made it to stick faster to his backe, for the more the Winde encreased the closer his cloake clapt to his body: then the Sunne, shining with his hot beams, began to warm this gentleman, who waxing somewhat faint in his faire weather, did not only put off his cloake but his coate, which the Wynde perceiving, yeelded the conquest to the Sunne."

Always remember that men are more easily led than driven, and that in any case it is better to guide than to coerce. "What thou wilt, thou rather shalt enforce it with thy smile, than hew to't with thy sword."1

Sydney Smith used to say of Francis Horner, who, without holding any high office, exercised a remarkable personal influence in the Councils of the Nation, that he had the Ten Commandments stamped upon his countenance.

Try to meet the wishes of others as far as you rightly and wisely can; but do not be afraid to say "No." Anybody can say "Yes," though it is not every one who can say "Yes" pleasantly; but it is far more difficult to say "No." Many a man has been ruined because he could not do so. Plutarch tells us that the inhabitants of Asia Minor came to be vassals only for not having been able to pronounce one syllable, which is "No." And if in the Conduct of Life it is essential to say "No," it is scarcely less necessary to be able to say it pleasantly. We ought always to endeavour that everybody with whom we have any transactions should feel that it is a pleasure to do business with us and should wish to come again. Business is a matter of sentiment and feeling far more than many suppose; every one likes being treated with kindness and courtesy, and a frank pleasant manner will often clench a bargain more effectually than a half per cent.

Almost any one may make himself pleasant if he wishes. "The desire of pleasing is at least half the art of doing it:" and, on the other hand, no one will please others who does not desire to do so. If you do not acquire this great gift while you are young, you will find it much more difficult afterwards. Many a man has owed his outward success in life far more to good manners than to any solid merit; while, on the other hand, many a worthy man, with a good heart and kind intentions, makes enemies merely by the roughness of his manner. To be able to please is, moreover, itself a great pleasure. Try it, and you will not be disappointed.

Be wary and keep cool. A cool head is as necessary as a warm heart. In any negotiations, steadiness and coolness are invaluable; while they will often carry you in safety through times of danger and difficulty.

If you come across others less clever than you are, you have no right to look down on them. There is nothing more to be proud of in inheriting great ability, than a great estate. The only credit in either case is if they are used well. Moreover, many a man is much cleverer than he seems. It is far more easy to read books than men. In this the eyes are a great guide. "When the eyes say one thing and the tongue another, a practised man relies on the language of the first."

Do not trust too much to professions of extreme goodwill. Men do not fall in love with men, nor women with women, at first sight. If a comparative stranger protests and promises too much, do not place implicit confidence in what he says. If not insincere, he probably says more than he means, and perhaps wants something himself from you. Do not therefore believe that every one is a friend, merely because he professes to be so; nor assume too lightly that any one is an enemy.

Argument is always a little dangerous. If often leads to coolness and misunderstandings. You may gain your argument and lose your friend, which is probably a bad bargain. If you must argue, admit all you can, but try and show that some point has been overlooked. Very few people know when they have had the worst of an argument, and if they do, they do not like it. Moreover, if they know they are beaten, it does not follow that they are convinced. Indeed it is perhaps hardly going too far to say that it is very little use trying to convince any one by argument. State your case as clearly and concisely as possible, and if you shake his confidence in his own opinion it is as much as you can expect. It is the first step gained.

Do not expect too much attention when you are young. Sit, listen, and look on. Bystanders proverbially see most of the game; and you can notice what is going on just as well, if not better, when you are not noticed yourself. It is almost as if you possessed a cap of invisibility. To save themselves the trouble of thinking, which is to most people very irksome, men will often take you at your own valuation. "*On ne vault dans ce monde*," says La Bruyère, "*que ce que l'on veult valoir.*" Do not

Charles Darwin, the Copley Medal, and the Rise of Naturalism

make enemies for yourself; you can make nothing worse. "Answer not a fool according to his folly, lest thou also be like unto him."

Remember that "a soft answer turneth away wrath;" but even an angry answer is less foolish than a sneer: nine men out of ten would rather be abused, or even injured, than laughed at. They will forget almost anything sooner than be made ridiculous. "It is pleasanter to be deceived than to be undeceived." Trasilaus, an Athenian, went mad, and thought that all the ships in the Piræus belonged to him, but having been cured by Crito, he complained bitterly that he had been robbed. It is folly, says Lord Chesterfield, "to lose a friend for a jest: but, in my mind, it is not much less degree of folly, to make an enemy of an indifferent and neutral person for the sake of a *bon-mot*."

Do not be too ready to suspect a slight, or think you are being laughed at—to say with Scrub in the Stratagem, "I am sure they talked of me, for they laughed consumedly." On the other hand, if you are laughed at, try to rise above it. If you can join in heartily, you will turn the tables and gain rather than lose. Every one likes a man who can enjoy a laugh at his own expense - and justly so, for it shows good-humour and good-sense. If you laugh at yourself, other people will not laugh at you. Have the courage of your opinions. You must expect to be laughed at sometimes, and it will do you no harm. There is nothing ridiculous in seeming to be what you really are, but a good deal in affecting to be what you are not. People often distress themselves, get angry, and drift into a coolness with others, for some quite imaginary grievance.

Be frank, and yet reserved. Do not talk much about yourself; neither of yourself, for yourself, nor against yourself: but let other people talk about themselves, as much as they will. If they do so it is because they like it, and they will think all the better of you for listening to them. At any rate do not show a man, unless it is your duty, that you think he is a fool or a blockhead. If you do, he has good reason to complain. You may be wrong in your judgment; he will, and with some justice, form the same opinion of you. Burke once said that he could not draw an indictment against a nation, and it is very unwise as well as unjust to attack any class or profession. Individuals often forget and forgive, but Societies never do. Moreover, even individuals will forgive an injury much more readily than an insult. Nothing rankles so much as being made ridiculous. You will never gain your object by putting people out of humour, or making them look ridiculous.

Goethe in his "Conversations with Eckermann" commended our countrymen. Their entrance and bearing in Society, he said, were so confident and quiet that one would think they were everywhere the masters, and the whole world belonged to them. Eckermann replied that surely young Englishmen were no cleverer, better educated, or better hearted than young Germans. "That is not the point," said

Goethe; "their superiority does not lie in such things, neither does it lie in their birth and fortune: it lies precisely in their having the courage to be what nature made them. There is no halfness about them. They are complete men. Sometimes complete fools, also, that I heartily admit; but even that is something, and has its weight."

In any business or negotiations, be patient. Many a man would rather you heard his story than granted his request: many an opponent has been tired out. Above all, never lose your temper, and if you do, at any rate hold your tongue, and try not to show it. "Cease from anger, and forsake wrath: Fret not thyself in any wise to do evil." For "A softer answer turneth away wrath: But grievous words stir up anger." Never intrude where you are not wanted. There is plenty of room elsewhere. "Have I not three kingdoms?" said King James to the fly, "and yet thou must needs fly in my eye." Some people seem to have a knack of saying the wrong thing, of alluding to any subject which revives sad memories, or rouses differences of opinion.

No branch of Science is more useful than the knowledge of Men. It is of the utmost importance to be able to decide wisely, not only to know whom you can trust, and whom you cannot, but how far, and in what, you can trust them. This is by no means easy. It is most important to choose well those who are to work with you, and under you; to put the square man in the square hole, and the round man in the round hole. "If you suspect a man, do not employ him: if you employ him, do not suspect him." Those who trust are oftener right than those who mistrust. Confidence should be complete, but not blind. Merlin lost his life, wise as he was, for imprudently yielding to Vivien's appeal to trust her, "all in all or not at all." Be always discreet.

Keep your own counsel. If you do not keep it for yourself, you cannot expect others to keep it for you. "The mouth of a wise man is in his heart; the heart of a fool is in his mouth, for what he knoweth or thinketh he uttereth." Use your head. Consult your reason. It is not infallible, but you will be less likely to err if you do so.

Speech is, or ought to be silvern, but silence is golden. Many people talk, not because they have anything to say, but for the mere love of talking. Talking should be an exercise of the brain, rather than of the tongue. Talkativeness, the love of talking for talking's sake, is almost fatal to success. Men are "plainly hurried on, in the heat of their talk, to say quite different things from what they first intended, and which they afterwards wish unsaid: or improper things, which they had no other end in saying, but only to find employment to their tongue. And this unrestrained volubility and wantonness in speech is the occasion of numberless evils and vexations in life. It begets resentment in him who is the subject of it; sows the seed of strife and dissension amongst others; and inflamed little disgusts and offences, which, if let alone, would wear away of themselves."

Never try to show your own superiority: few things annoy people more than being made to feel small. Do not be too positive in your statements. You may be wrong, however sure you feel. Memory plays us curious tricks, and both ears and eyes are sometimes deceived. Our prejudices, even the most cherished, may have no secure foundation. Moreover, even if you are right, you will lose nothing by disclaiming too great certainty. In action, again, never make too sure, and never throw away a chance. "There's many a slip 'twixt the cup and the lip." It has been said that everything comes to those who know how to wait; and when the opportunity does come, seize it. "He that wills not, when he may; when he will, he shall have nay." If you once let your opportunity go, you may never have another. "There is a tide in the affairs of men, Which taken at the flood, leads on to fortune: Omitted, all the voyage of their life is bound in shallows and in miseries. On such a full sea are we now afloat: And we must take the current when it serves, Or lose our venture." Be cautious, but not over-cautious; do not be too much afraid of making a mistake; "a man who never makes a mistake, will make nothing."

FRANCIS BACON'S *NOVUM ORGANUM*, 1620 (EXCERPTS)

[The following excerpts were adapted from James Spedding, Robert Leslie Ellis, and Douglas Denon Heath, trans., *Novum Organum*, in *The Works* [of Francis Bacon], vol. 8, Boston: Taggard and Thompson, 1863. Editor's notes appear in brackets [ ].]

Preface

. . . Now my method, though hard to practice, is easy to explain; and it is this. I propose to establish progressive stages of certainty. The evidence of the sense, helped and guarded by a certain process of correction, I retain. But the mental operation which follows the act of sense I for the most part reject; and instead of it I open and lay out a new and certain path for the mind to proceed in, starting directly from the simple sensuous perception. The necessity of this was felt, no doubt, by those who attributed so much importance to logic, showing thereby that they were in search of helps for the understanding, and had no confidence in the native and spontaneous process of the mind. But this remedy comes too late to do any good, when the mind is already, through the daily intercourse and conversation of life, occupied with unsound doctrines and beset on all sides by vain imaginations. And therefore that art of logic, coming (as I said) too late to the rescue, and no way able to set matters right again, has had the effect of fixing errors rather than disclosing truth. There remains but one course for the recovery of a sound and healthy condition — namely, that the entire work of the understanding be commenced afresh, and the mind itself be from the very outset not left to take its own course, but guided at every step; and the business be done as if by machinery. . . .

[In this preface, Bacon introduces a new method for scientific inquiry, which he calls the "inductive method." There are a few particularly noteworthy points. First, he proposes that his method will "establish **progressive stages of certainty**." This is revolutionary among methods of inquiry. The deductive method favored by Bacon's predecessors aims at absolute certainty. Indeed, a deductive method is incapable of providing answers of various degrees of certainty. Bacon rightly believes his method cannot lead to absolute certainty. Second, Bacon states that his method starts with nothing but "simple sensuous perception." This denies that the imagination, speculation, or dependence on observed phenomenon can be utilized in a scientific inquiry. Third, the method relies on tools that operate independently of the mind, or "machinery." In this way, the errors introduced by human assumption are reduced.]

Aphorisms

I.

Man, being the servant and interpreter of Nature, can do and understand so much and so much only as he has observed in fact or in thought of the course of nature. Beyond this he neither knows anything nor can do anything.

[There has always been dispute about what sorts of ideas count as legitimate knowledge, and this was true in Bacon's time more than in most others. Here, he is stating that knowledge results only from observation of nature. All other ideas cannot be counted as knowledge, and as a result, it is unlikely that these ideas will lead to any productive activity. Besides this, he is also claiming that any theory that depends on speculation, even to a small degree, is not legitimate knowledge.]

II.

Neither the naked hand nor the understanding left to itself can effect much. It is by instruments and helps that the work is done, which are as much wanted for the understanding as for the hand. And as the instruments of the hand either give motion or guide it, so the instruments of the mind supply either suggestions for the understanding or cautions.

[Humans are Prone to Anticipations, or Speculations]

XXVII.

Anticipations are a ground sufficiently firm for consent, for even if men went mad all after the same fashion, they might agree one with another well enough.

[The term "anticipations" is nearly synonymous with "ideas resulting from speculation." Even though he states that anticipations are sufficient for consent, this is still a criticism. Consent does not imply truth, but could very likely mean that a large population can be incorrect in the same way.]

XXVIII.

For the winning of assent, indeed, anticipations are far more powerful than interpretations, because being collected from a few instances, and those for the most part of familiar occurrence, they straightway touch the understanding and fill the imagination; whereas interpretations, on the other hand, being gathered here and there from very various and widely dispersed facts, cannot suddenly strike the understanding; and therefore they must needs, in respect of the opinions of the time, seem harsh and out of tune, much as the mysteries of faith do.

[Some Motivations for our Anticipations/Speculations]

XLV.

The human understanding is of its own nature prone to suppose the existence of more order and regularity in the world than it finds. And though there be many things in nature which are singular and unmatched, yet it devises for them parallels and conjugates and relatives which do not exist. Hence the fiction that all celestial bodies move in perfect circles, spirals and dragons being (except in name) utterly rejected. Hence too the element of fire with its orb is brought in, to make up the square with the other three which the sense perceives. Hence also the ratio of density of the so-called elements is arbitrarily fixed at ten to one. And so on of other dreams. And these fancies affect not dogmas only, but simple notions also.

[Bacon is describing one reason why humans often "anticipate" the truth of ideas prior to discovering them through observation. He is claiming that humans are inherently prejudiced to believe that the world is orderly, and that the world makes logical sense. Therefore, he continues, when presented with a large number of observations, rather than discovering through legitimate means some way to make sense of our observations, we will use speculation to create illegitimate theories to make sense of the world. We would rather have a false explanation than no explanation at all.]

XLVI.

The human understanding when it has once adopted an opinion (either as being the received opinion or as being agreeable to itself) draws all things else to support and agree with it. And though there be a greater number and weight of instances to be found on the other side, yet these it either neglects and despises, or else by some distinction sets aside and rejects, in order that by this great and pernicious predetermination the authority of its former conclusions may remain inviolate. And therefore it was a good answer that was made by one who, when they showed him hanging in a temple a picture of those who had paid their vows as having escaped shipwreck, and would have him say whether he did not now acknowledge the power of the gods —"Aye," asked he again, "but where are they painted that were drowned after their vows?" And such is the way of all superstition, whether in astrology, dreams, omens, divine judgments, or the like; wherein men, having a delight in such vanities, mark the events where they are fulfilled, but where they fail, though this happen much oftener, neglect and pass them by. But with far more subtlety does this mischief insinuate itself into philosophy and the sciences; in which the first conclusion colors and brings into conformity with itself all that come after, though far sounder and better. Besides, independently of that delight and vanity which I have described, it is the peculiar and perpetual error of the human intellect to be more moved and excited by affirmatives than by negatives; whereas it ought properly to hold itself indifferently disposed toward both alike. Indeed, in the establishment of any true axiom, the negative instance is the more forcible of the two.

[In XLVI, Bacon is describing what is now called "confirmation bias." This is another reason that humans often rely on anticipations rather than legitimate knowledge. Once an explanation or theory has been accepted as true, we are more inclined to find and accept those observations (even legitimate observations) that support the explanation or theory, and less likely to find and accept those observations which would contradict the explanation or theory.]

LI.

The human understanding is of its own nature prone to abstractions and gives a substance and reality to things which are fleeting. But to resolve nature into abstractions is less to our purpose than to dissect her into parts; as did the school of Democritus, which went further into nature than the rest. Matter rather than forms should be the object of our attention, its configurations and changes of configuration, and simple action, and law of action or motion; for forms are figments of the human mind, unless you will call those laws of action forms.

[It is easier to understand something that does not change than to understand something that does change. Therefore, humans are prone to understand things as unchanging rather than changing. Effectively, the human mind is prone to seeing things not as they really are, but rather as an unchanging abstraction. Therefore, our sciences are prone to study fictions rather than realities. Of particular relevance to the Darwin controversy, Aphorism LXVI states, " . . . when man contemplates nature working freely, he meets with different species of things, of animals, of plants, of minerals; whence he readily passes into the opinion that there are in nature certain primary forms which nature intends to educe, and that the remaining variety proceeds from hindrances and aberrations of nature in the fulfillment of her work, or from the collision of different species and the transplanting of one into another . . . "]

[Notes on the use of induction to new discoveries]

[Bacon describes the method of induction as a very slow, careful transcendence from knowledge of particulars (observations of single things) toward, eventually, general laws, or what he calls "axioms." In CII, he warns that the data collected from the "army of particulars" must be organized into tables, so that our intellect, with its proneness to anticipation, does not accidentally infer things that it has no right to. In CIII, Bacon admits that after transcending to the axioms, the axioms can then be used to point the way to new particulars, allowing us to once again transcend to new axioms (for example, using Newton's Laws of Motion to discover new things about the world, such as figuring out the necessary velocity of a space shuttle as it prepares for re-entry into the atmosphere). In CIV and CV, Bacon further describes the ascent from particulars to axioms, somewhat distinguishing between lower, middle, and higher axioms. He further warns the reader of the danger of letting our ideas leap and fly toward axioms and directs that we should add weights to insure a slow and steady ascent.]

CII.

Moreover, since there is so great a number and army of particulars, and that army so scattered and dispersed as to distract and confound the understanding, little is to be hoped for from the skirmishings and slight attacks and desultory movements of the intellect, unless all the particulars which pertain to the subject of inquiry shall, by means of Tables of Discovery, apt, well arranged, and, as it were, animate, be drawn up and marshaled; and the mind be set to work upon the helps duly prepared and digested which these tables supply.

CIII.

But after this store of particulars has been set out duly and in order before our eyes, we are not to pass at once to the investigation and discovery of new particulars or works; or at any rate if we do so we must not stop there. For although I do not deny that when all the experiments of all the arts shall have been collected and digested, and brought within one man's knowledge and judgment, the mere transferring of the experiments of one art to others may lead, by means of that experience which I term literate, to the discovery of many new things of service to the life and state of man, yet it is no great matter that can be hoped from that; but from the new light of axioms, which having been educed from those particulars by a certain method and rule, shall in their turn point out the way again to new particulars, greater things may be looked for. For our road does not lie on a level, but ascends and descends; first ascending to axioms, then descending to works.

CIV.

The understanding must not, however, be allowed to jump and fly from particulars to axioms remote and of almost the highest generality (such as the first principles, as they are called, of arts and things), and taking stand upon them as truths that cannot be shaken, proceed to prove and frame the middle axioms by reference to them;

which has been the practice hitherto, the understanding being not only carried that way by a natural impulse, but also by the use of syllogistic demonstration trained and inured to it. But then, and then only, may we hope well of the sciences when in a just scale of ascent, and by successive steps not interrupted or broken, we rise from particulars to lesser axioms; and then to middle axioms, one above the other; and last of all to the most general. For the lowest axioms differ but slightly from bare experience, while the highest and most general (which we now have) are notional and abstract and without solidity. But the middle are the true and solid and living axioms, on which depend the affairs and fortunes of men; and above them again, last of all, those which are indeed the most general; such, I mean, as are not abstract, but of which those intermediate axioms are really limitations.

The understanding must not therefore be supplied with wings, but rather hung with weights, to keep it from leaping and flying. Now this has never yet been done; when it is done, we may entertain better hopes of the sciences.

CV.

In establishing axioms, another form of induction must be devised than has hitherto been employed, and it must be used for proving and discovering not first principles (as they are called) only, but also the lesser axioms, and the middle, and indeed all. For the induction which proceeds by simple enumeration is childish; its conclusions are precarious and exposed to peril from a contradictory instance; and it generally decides on too small a number of facts, and on those only which are at hand. But the induction which is to be available for the discovery and demonstration of sciences and arts, must analyze nature by proper rejections and exclusions; and then, after a sufficient number of negatives, come to a conclusion on the affirmative instances — which has not yet been done or even attempted, save only by Plato, who does indeed employ this form of induction to a certain extent for the purpose of discussing definitions and ideas. But in order to furnish this induction or demonstration well and duly for its work, very many things are to be provided which no mortal has yet thought of; insomuch that greater labor will have to be spent in it than has hitherto been spent on the syllogism. And this induction must be used not only to discover axioms, but also in the formation of notions. And it is in this induction that our chief hope lies.

[On the Mixing of Theology and Science]

LXV.

But the corruption of philosophy by superstition and an admixture of theology is far more widely spread, and does the greatest harm, whether to entire systems or to their parts. For the human understanding is obnoxious to the influence of the imagination no less than to the influence of common notions. For the contentious and sophistical kind of philosophy ensnares the understanding; but this kind, being fanciful and tumid and half poetical, misleads it more by flattery. For there is in man an ambition of the understanding, no less than of the will, especially in high and lofty spirits.

Of this kind we have among the Greeks a striking example in Pythagoras, though he united with it a coarser and more cumbrous superstition; another in Plato and his school, more dangerous and subtle. It shows itself likewise in parts of other philosophies, in the introduction of abstract forms and final causes and first causes, with the omission in most cases of causes intermediate, and the like. Upon this point the greatest caution should be used. For nothing is so mischievous as the apotheosis of error; and it is a very plague of the understanding for vanity to become the object of veneration. Yet in this vanity some of the moderns have with extreme levity indulged so far as to attempt to found a system of natural philosophy on the first chapter of Genesis, on the book of Job, and other parts of the sacred writings, seeking for the dead among the living; which also makes the inhibition and repression of it the more important, because from this unwholesome mixture of things human and divine there arises not only a fantastic philosophy but also a heretical religion. Very meet it is therefore that we be sober-minded, and give to faith that only which is faith's.

WILLIAM PALEY, *NATURAL THEOLOGY; OR, EVIDENCES OF THE EXISTENCE AND ATTRIBUTES OF THE DEITY*

[The following version has been abridged and annotated from Paley, *Natural Theology, or, Evidences of the Existence and Attributes of the Deity Collected from the Appearances of Nature,* Boston: Gould and Lincoln, 1855, pp. 5–12, 44–45. Editor's notes appear in brackets [ ].]

State of the Argument

IN crossing a heath, suppose I pitched my foot against a stone, and were asked how the stone came to be there; I might possibly answer, that, for any thing I knew to the contrary, it had lain there for ever: nor would it perhaps be very easy to show the absurdity of this answer. But suppose I had found a watch upon the ground, and it should be inquired how the watch happened to be in that place; I should hardly think of the answer which I had before given, that, for any thing I knew, the watch might have always been there. Yet why should not this answer serve for the watch as well as for the stone? why is it not as admissible in the second case, as in the first? For this reason, and for no other, viz. that, when we come to inspect the watch, we perceive (what we could not discover in the stone) that its several parts are framed and put together for a purpose, *e. g.* that they are so formed and adjusted as to produce motion, and that motion so regulated as to point out the hour of the day; that, if the different parts had been differently shaped from what they are, of a different size from what they are, or placed after any other manner, or in any other order, than that in which they are placed, either no motion at all would have been carried on in the machine, or none which would have answered the use that is now served by it.

This mechanism being observed (it requires indeed an examination of the instrument, and perhaps some previous knowledge of the subject, to perceive and understand it; but being once, as we have said, observed and understood), the inference, we think, is inevitable, that the watch must have had a maker: that there must have existed, at some time, and at some place or other, an artificer or artificers who formed it for the purpose which we find it actually to answer; who comprehended its construction, and designed its use.

Nor would it, I apprehend, weaken the conclusion, that we had never seen a watch made; that we had never known an artist capable of making one; that we were altogether incapable of executing such a piece of workmanship ourselves, or of understanding in what manner it was performed; all this being no more than what is true of some exquisite remains of ancient art, of some lost arts, and, to the generality of mankind, of the more curious productions of modern manufacture.

[handwritten note: DK curiosity around specified watch dev't →]

Neither, secondly, would it invalidate our conclusion, that the watch sometimes went wrong, or that it seldom went exactly right. The purpose of the machinery, the design, and the designer, might be evident, and in the case supposed would be evident, in whatever way we accounted for the irregularity of the movement, or whether we could account for it or not. It is not necessary that a machine be perfect, in order to show with what design it was made: still less necessary, where the only question is, whether it were made with any design at all.

THIS is atheism: for every indication of contrivance, every manifestation of design, which existed in the watch, exists in the works of nature; with the difference, on the side of nature, of being greater and more, and that in a degree which exceeds all computation. I mean that the contrivances of nature surpass the contrivances of art, in the complexity, subtlety, and curiosity of the mechanism; and still more, if possible, do they go beyond them in number and variety; yet, in a multitude of cases, are not less evidently mechanical, not less evidently contrivances, not less evidently accommodated to their end, or suited to their office, than are the most perfect productions of human ingenuity.

I know no better method of introducing so large a subject, than that of comparing a single thing with a single thing; an eye, for example, with a telescope. As far as the examination of the instrument goes, there is precisely the same proof that the eye was made for vision, as there is that the telescope was made for assisting it. They are made upon the same principles; both being adjusted to the laws by which the transmission and refraction of rays of light are regulated. I speak not of the origin of the laws themselves; but such laws being fixed, the construction, in both cases, is adapted to them. For instance; these laws require, in order to produce the same effect, that the rays of light, in passing from water into the eye, should be refracted by a more convex surface, than when it passes out of air into the eye. Accordingly we find that the eye of a fish, in that part of it called the crystalline lens, is much rounder than the eye of terrestrial animals. What plainer manifestation of design can there be than this difference? What could a mathematical-instrument-maker have done more, to show his knowledge of his principle, his application of that knowledge, his suiting of his means to his end; I will not say to display the compass or excellence of his skill and art, for in these all comparison is indecorous, but to testify counsel, choice, consideration, purpose?

To some it may appear a difference sufficient to destroy all similitude between the eye and the telescope, that the one is a perceiving organ, the other an unperceiving instrument. The fact is, that they are both instruments. And, as to the mechanism, at least as to mechanism being employed, and even as to the kind of it, this circumstance varies not the analogy at all. For observe, what the constitution of the eye is . . . [details re. retina, pupil, lens].

Charles Darwin, the Copley Medal, and the Rise of Naturalism

Can any thing be more decisive of contrivance than this is? The most secret laws of optics must have been known to the author of a structure endowed with such a capacity of change. It is as though an optician, when he had a nearer object to view, should *rectify* his instrument by putting in another glass, at the same time drawing out also his tube to a different length.

Observe a new-born child first lifting up its eyelids. What does the opening of the curtain discover? The anterior part of two pellucid globes, which, when they come to be examined, are found to be constructed upon strict optical principles; the self-same principles upon which we ourselves construct optical instruments. We find them perfect for the purpose of forming an image by refraction; composed of parts executing different offices: one part having fulfilled its office upon the pencil of light, delivering it over to the action of another part; that to a third, and so onward: the progressive action depending for its success upon the nicest and minutest adjustment of the parts concerned; yet, these parts so in fact adjusted, as to produce, not by a simple action or effect, but by a combination of actions and effects, the result which is ultimately wanted. And forasmuch as this organ would have to operate under different circumstances, with strong degrees of light, and with weak degrees, upon near objects, and upon remote ones, and these differences demanded, according to the laws by which the transmission of light is regulated, a corresponding diversity of structure; that the aperture, for example, through which the light passes, should be larger or less; the lenses rounder or flatter, or that their distance from the tablet, upon which the picture is delineated, should be shortened or lengthened: this, I say, being the case and the difficulty, to which the eye was to be adapted, we find its several parts capable of being occasionally changed, and a most artificial apparatus provided to produce that change. This is far beyond the common regulator of a watch, which requires the touch of a foreign hand to set it: but it is not altogether unlike Harrison's contrivance for making a watch regulate itself, by inserting within it a machinery, which, by the artful use of the different expansion of metals, preserves the equability of the motion under all the various temperatures of heat and cold in which the instrument may happen to be placed. The ingenuity of this last contrivance has been justly praised. Shall, therefore, a structure which differs from it, chiefly by surpassing it, be accounted no contrivance at all? or, if it be a contrivance, that it is without a contriver!

But this, though much, is not the whole; by different species of animals the faculty we are describing is possessed, in degrees suited to the different range of vision which their mode of life, and of procuring their food, requires. [examples of various animals' and plants' particularly well suited characteristics.]

[Additional arguments against the existence of the natural world as pure happenstance]

Others have chosen to refer every thing to a *principle of order* in nature. A principle of order is the word: but what is meant by a principle of order, as different from an intelligent Creator, has not been explained either by definition or example: and, without such explanation, it should seem to be a mere substitution of words for reasons, names for causes. Order itself is only the adaptation of means to an end: a principle of order therefore can only signify the mind and intention which so adapts them. Or, were it capable of being explained in any other sense, is there any experience, any analogy, to sustain it? Was a watch ever produced by a principle of order? and why might not a watch be so produced, as well as an eye?

Furthermore, a principle of order, acting blindly, and without choice, is negatived, by the observation, that order is not universal; which it would be, if it issued from a constant and necessary principle: nor indiscriminate, which it would be, if it issued from an unintelligent principle. Where order is wanted, there we find it; where order is not wanted, *i. e.* where, if it prevailed, it would be useless, there we do not find it. In the structure of the eye (for we adhere to our example), in the figure and position of its several parts, the most exact order is maintained. In the forms of rocks and mountains, in the lines which bound the coasts of continents and islands, in the shape of bays and promontories, no order whatever is perceived, because it would have been superfluous. No useful purpose would have arisen from moulding rocks and mountains into regular solids, bounding the channel of the ocean by geometrical curves; or from the map of the world, resembling a table of diagrams in Euclid's Elements, or Simpson's Conic Sections.

The Argument Cumulative

WERE there no example in the world, of contrivance, except that of the *eye*, it would be alone sufficient to support the conclusion which we draw from it, as to the necessity of an intelligent Creator. It could never be got rid of; because it could not be accounted for by any other supposition, which did not contradict all the principles we possess of knowledge; the principles, according to which, things do, as often as they can be brought to the test of experience, turn out to be true or false. Its coats and humours, constructed, as the lenses of a telescope are constructed, for the refraction of rays of light to a point, which forms the proper action of the organ; the provision in its muscular tendons for turning its pupil to the object, similar to that which is given to the telescope by screws, and upon which power of direction in the eye, the exercise of its office as an optical instrument depends; the further provision for its defence, for its constant lubricity and moisture, which we see in its socket and its lids, in its gland for the secretion of the matter of tears, its outlet or communication with the nose for carrying off the liquid after the eye is washed with it; these provisions compose altogether an apparatus, a system of parts, a preparation of means, so manifest in their design, so exquisite in their contrivance, so successful in their issue, so precious, and so infinitely beneficial in their use, as, in my opinion, to bear down all doubt that can be raised upon the subject. And what

I wish, under the title of the present chapter, to observe is, that if other parts of nature were inaccessible to our inquiries, or even if other parts of nature presented nothing to our examination but disorder and confusion, the validity of this example would remain the same. If there were but one watch in the world, it would not be less certain that it had a maker. If we had never in our lives seen any but one single kind of hydraulic machine, yet, if of that one kind we understood the mechanism and use, we should be as perfectly assured that it proceeded from the hand, and thought, and skill of a workman, as if we visited a museum of the arts, and saw collected there twenty different kinds of machines for drawing water, or a thousand different kinds for other purposes. Of this point, each machine is a proof, independently of all the rest. So it is with the evidences of a Divine agency. The proof is not a conclusion which lies at the end of a chain of reasoning, of which chain each instance of contrivance is only a link, and of which, if one link fail, the whole falls; but it is an argument separately supplied by every separate example. An error in stating an example, affects only that example. The argument is cumulative, in the fullest sense of that term. The eye proves it without the ear; the ear without the eye. The proof in each example is complete; for when the design of the part, and the conduciveness of its structure to that design is shown, the mind may set itself at rest; no future consideration can detract any thing from the force of the example.

CHARLES KINGSLEY, "A NATION'S GRIEF FOR A NATION'S LOSS"

[Below is the text of Charles Kingsley's sermon preached at Eversley Church, December 22, "A Sermon on the Death of His Royal Highness the Prince Consort," London: Parker, 1862.]

Text: "Know ye not that there is a prince and a great man fallen this day in Israel?"

Explication: The words of the Hebrew King, my brethren, spoken in the far-off ages, seem to rise almost spontaneously to our lips today. The solemn funeral procession, the weeping monarch—whose deep sobs found a sympathizing response in those of his people—as the remains of a prince and a general, who seems to have inspired a more than usual degree of love and affection, were slowly borne in Hebron to their resting-place—the whole scene, as described in the short and touching language of Holy Scripture, rises before our eyes on an occasion, which, like this, so strongly recalls it in all its grand simplicity and vivid truthfulness.

. . . A heavy and unexpected blow has fallen on our Queen, and on our land. . . . God calls on her to sustain the heaviest bereavement which can befall woman—the loss of a faithful and well-loved husband. Last Sunday morning, as men and women were quietly assembling in God's House of Prayer, the sad rumour began to steal from mouth to mouth that in the still watches of the night preceding, in the chambers of the royal Castle of England, the soul of Albert, Prince Consort, had passed away. . . . The man of all others occupied the highest position in this island, the husband of our Queen, the father of our future King, in the very prime of life, in the very vigour of robust manhood, placed at the very height of human happiness, surrounded by a loving wife and affectionate children, with the prospect stretching far before him of an useful and respected old age, suddenly—and without the faintest indication of approaching disease and death—was stricken down and taken away from among us. All that human science and skill could effect—all the alleviations which unlimited wealth could procure—were there, and all failed. "The appointed time" had come and the soul returned to God Who gave it. . . .

Application: And now, brethren, a few thoughts for ourselves. What are the practical lessons God would have us each learn from the sad event? They are, I think, very plain and obvious. They are the old, old lessons which God is ever teaching, and man is ever forgetting—the uncertainty of life, the certainty of death! A few days ago, and, perhaps, there was no man of his age in England, for whom a lengthened life might have been more confidently predicted than for the Prince who has just been cut off so suddenly. But death, my brethren, is no respecter of persons, times, or places. He knocks at the palace gates not less loudly than at the door of the humblest hovel. He carries off the old in their weakness, and the young in their strength. No human science or skill is able to arrest his approach—no strength of human love can shelter from his fatal darts. These, my brethren, may seem old and time-worn truths, but are they the less true for that? Have we realized them for ourselves? Do we live in the faith of them? A few years more, at the most—it may be much, much less, God only knows!—and death, which we are so apt to speak of as though it affected others only, will have come home as a present reality to each one of us. We, too, shall very, very soon be lying on our death-bed. Our weeping friends and relations will stand around, but their love cannot interpose to save us; physicians will, perhaps, still offer their medicines, but all their skill will not avail to rescue us; prayers will, perhaps, ascend, with faltering lips, to Heaven, but the will of the Most High has ordained otherwise. None may descend with us into that dark valley: by ourselves we must plant our foot in the waters of the icy river—we must die alone. Then the soul will be brought face to face with her God, and the false and flimsy veils wherewith she has endeavoured to clothe her sins from her own sight, and that of her God, shall drop off one by one, and leave her desolate! Then the voice of Conscience, so long stifled, will once more assert her right, and make herself heard to the dismay of the sinner's heart! Then, for the first time, shall we see ourselves as we are seen by God. O what an hour of dread decision and awful heart-searching will that be! . . .

Exhortation: My brethren, this is no mere possibility I am setting before you, wherewith to affright your imagination and insult your common sense. It is a simple, sober certainty. That hour of death must, and will, be lived through by every man, woman, and child in this church, unless the Lord should first come to judge the quick and dead. Are you ready for that hour to come? Are you—in a word—prepared for death? . . .

. . . If your conscience tells you that you have not yet repented of the sins of your past life, that you are still living in willful sin, O while it is yet called the day, before the night of death come upon you (as come it *must*, and *may* at any time), while yet the longsuffering of God waiteth, return unto the Lord Who loves you, seek the embraces of the Good Shepherd Who has given his life for you. . . . Only confess, and all will be forgiven. . . . Think what I call you to give up? A few cherished sins and self-indulgences, which give you no real rest or peace. Think to what I invite you? The sight of God, the presence of Christ, the society of the saints, to be kings and priests for evermore. O who can doubt for a moment? Who will barter his eternal birthright, for this world's mess of pottage? Ad if you have, my dear brethren, made your peace with God, if you have forsaken your sins, and are daily endeavouring to do His will on earth as it is done in heaven, think what that solemn hour of death will be to you then. All its real terror will be past and gone. "Thou shalt keep him in perfect peace whose mind is stayed on Thee." And when the dark shadows of the valley fall thickest round us, and the rolling stream of the river of death threaten to go even over our soul, the Hand of Jesus will sustain us, in the passage, and His voice shall be heard speaking to our inmost heart—"it is I; be not afraid."

SAMPLE PRAYERS FROM THE BOOK OF COMMON PRAYER

[From *The Book of Common Prayer and Administration of the Sacraments,* London: Bickers and Bush, 1863. The 1662 version of the prayer book remained in use with few changes (except to adapt prayers to the current political scene) until the twentieth century. Editor's notes appear in brackets [ ].]

Morning Prayer

Almighty and most merciful Father,

We have erred and strayed from thy ways like lost sheep. We have followed too much the devices and desires of our own hearts. We have offended against thy holy laws. We have left undone those things which we ought to have done, and we have done those things which we ought not to have done. And there is no health [goodness] in us. But thou, O Lord, have mercy upon us, miserable offenders. Spare thou them [Will you please spare those] O God, which confess their faults. Restore thou them that are penitent [sorry] according to thy promises declared unto mankind in Christ Jesus our Lord [save those who have faith that Christ died for our sins]. And grant, oh most merciful Father for his sake, that we may hereafter live a godly, righteous, and sober life. To the glory of thy holy Name. Amen.

A Prayer for the Queen's Majesty

Almighty God, the fountain of all goodness, King of kings, Lord of lords, the only Ruler of princes, who dost from thy throne behold all the dwellers upon earth,

Most heartily we beseech thee with thy favour to behold our most gracious Sovereign Lady, Queen Victoria, and so replenish her with the grace of thy Holy Spirit, that she may always incline to thy will and walk in thy way. Endue her plenteously with heavenly gifts. Grant her in health and wealth long to live. Strengthen her that she may vanquish and overcome all her enemies. And finally, after this life, she may attain everlasting joy and felicity. Through Jesus Christ our Lord. Amen.

Apostles' Creed

[A good reminder of the basic beliefs of protestant Christians and frequently recited together by the entire congregation.]

I believe in God the Father Almighty, maker of heaven and earth, and in Jesus Christ his only son our Lord, who was conceived by the Holy Ghost, born of the Virgin Mary, suffered under Pontius Pilate, was crucified, dead, and buried. He descended into hell. The third day he rose again from the dead. He ascended into heaven and sitteth on the right hand of God the Father Almighty. From thence he shall come to judge the quick and the dead. I believe in the Holy Ghost, the holy catholick [i.e. universal and unified] church, the communion of the Saints, the forgiveness of sins, the resurrection of the body, and the life everlasting. Amen.

Miscellaneous

Oh God, heavenly Father, whose gift it is that the rain doth fall, the earth is fruitful, beasts increase, and fishes do multiply,

Behold, we beseech thee, the afflictions of thy people and grant that the scarcity and dearth which we do now most justly suffer for our iniquity [have mercy though, because of our sinful nature, we deserve every terrible thing that happens to us], may through thy goodness be mercifully turned into cheapness and plenty [because you have mercy you may send us lots of what we need]; for the love of Jesus Christ our Lord, to whom with the and the Holy Ghost be all honour and glory, now and forever. Amen.

Oh God, merciful Father, who in the time of Elisha the prophet, didst suddenly in Samaria turn great scarcity and dearth into plenty and cheapness, have mercy upon us, that we, who are now for our sins punished with like adversity, may likewise find a seasonable relief. [We as that you] increase the fruits of the earth by thy heavenly benediction and grant that we, receiving thy bountiful liberality, may use the same to thy glory, the relief of those that are needy, and our own comfort, through Jesus Christ our Lord. Amen.

Hear us, Almighty and most merciful God and Saviour,

Extend thy accustomed goodness to this thy servant   who is grieved with sickness. Sanctify, we beseech thee, this thy fatherly correction to him; that the sense of his weak-ness may add strength to his faith, and seriousness to his repentance. That, if it shall be thy good pleasure to restore him to his former health, he may lead the residue of his life in thy fear, and to thy glory. Or else, give him grace so to take thy visitation, that, after this painful life ended, he may dwell with thee in life everlasting, through Jesus Christ our Lord. Amen.

[Note: Many passages from William Paley's *Natural Theology* (excerpted above) and other scientific writings lauding the wonders of God's universe may be easily adapted into an appropriate format for celebratory prayer. Use the prayers above as templates that can be altered for various occasions.]

SONG LYRICS: "GOD SAVE THE QUEEN"

[Tune: "My Country 'tis of Thee"]

God save our gracious Queen
Long live our noble Queen,
God save the Queen:
Send her victorious,
Happy and glorious,
Long to reign over us:
God save the Queen.

O Lord, our God, arise,
Scatter thine enemies,
And make them fall:
Confound their politics,
Frustrate their knavish tricks,
On thee our hopes we fix:
God save us all.

Thy choicest gifts in store,
On her be pleased to pour;
Long may she reign:
May she defend our laws,
And ever give us cause
To sing with heart and voice
God save the Queen.

SONG LYRICS: "ALL THINGS BRIGHT AND BEAUTIFUL"

By Cecil Alexander, 1848

Refrain

All things bright and beautiful,
All creatures great and small,
All things wise and wonderful:
The Lord God made them all.

Each little flower that opens,
Each little bird that sings,
He made their glowing colors,
He made their tiny wings.

1. Refrain

The rich man in his castle,
The poor man at his gate,
He made them, high or lowly,
And ordered their estate.

2. Refrain

The purple headed mountains,
The river running by,
The sunset and the morning
That brightens up the sky.

3. Refrain

The cold wind in the winter,
The pleasant summer sun,
The ripe fruits in the garden,
He made them every one.

4. Refrain

The tall trees in the greenwood,
The meadows where we play,
The rushes by the water,
To gather every day.

5. Refrain

He gave us eyes to see them,
And lips that we might tell
How great is God Almighty,
Who has made all things well.

6. Refrain

APPENDIX C. ADDITIONAL SOURCES

Secondary sources, including our narrative, should be consulted for background information and historical perspective on the events surrounding the Royal Society's debates.

Bartholomew, M. J. "The Award of the Copley Medal to Charles Darwin." *Notes and Records of the Royal Society of London* 30, no. 2 (1976): 209-218.

Barton, R. "'Huxley, Lubbock, and Half a Dozen Others': Professionals and Gentlemen in the Formation of the X Club, 1851-1864." *Isis* 89, no.3 (1998): 410-444.

Bellon, R. "Joseph Dalton Hooker's Ideals for a Professional Man of Science." *Journal of the History of Biology* 34 (2001): 51-82.

Brock, W.H. "The Selection of the Authors of the Bridgewater Treatises." *Notes and Records of the Royal Society of London* 21, no. 2 (1966): 162-179.

Burkhardt, F. H. "Darwin and the Copley Medal." *Proceedings of the American Philosophical Society* 145 (2001): 510-518.

Darwin, Charles. *The Correspondence of Charles Darwin.* Edited by Frederick Burkhardt, et al. 12 vols. Cambridge: Cambridge University Press, 1985-2001.

Cannon, W. F. "Scientists and Broad Churchmen: An Early Victorian Intellectual Network." *Journal of British Studies* 4, no.1 (1964): 65-88.

Chadwick, O. *The Secularization of the European Mind in the Nineteenth Century.* Cambridge: Cambridge University Press, 1975.

Crosland, M. "Explicit Qualifications as a Criterion for Membership of the Royal Society: A Historical Review." *Notes and Records of the Royal Society of London* 37, no. 2 (1983): 167-187.

Desmond, A. "Redefining the X Axis: 'Professionals,' 'Amateurs' and the Making of Mid-Victorian Biology—A Progress Report." *Journal of the History of Biology* 34 (2001): 3-50.

Francis, M. "The Origins of *Essays and Reviews:* An Interpretation of Mark Pattison in the 1850s." *Historical Journal* 17, no. 4 (1974): 797-811.

Hall, M. B. "The Royal Society in Thomas Henry Huxley's Time." *Notes and Records of the Royal Society of London* 38, no. 2 (1984): 153-158.

Herschel, J.F.W. *Physical Geography of the Globe.* Edinburgh: Adam and Charles Black, 1861.

Hull, David L. *Darwin and his Critics: The Reception of Darwin's Theory of Evolution by the Scientific Community.* Cambridge: Harvard University Press, 1973.

MacLeod, R. M. "The X-Club a Social Network of Science in Late-Victorian England." *Notes and Records of the Royal Society of London* 24, no. 2 (1970): 305-322.

Mandelbaum, M. "Darwin's Religious Views." *Journal of the History of Ideas* 19 (1958): 363-378.

Mayr, E. "The Philosophical Foundations of Darwinism." *Proceedings of the American Philosophical Society* 145 (2001): 488-495.

Mill, J. S. *A System of Logic, Ratiocinative and Inductive: Being a Connected View of the Principles of Evidence, and the Methods of Scientific Investigation,* 1843. Reprinted in vols VII and VIII of J.M. Robson, ed., *The Collected Works of John Stuart Mill,* 33 vols. Toronto: University of Toronto Press, 1963-91.

Ryan, F. X., Hoeveler, J. D., & Largent, M. *Darwinism and Theology in America: 1850-1930.* Dorset UK: Thoemmes Press, 2002.

Sedgwick, A. "Objections to Mr. Darwin's Theory of the Origin of Species." *The Spectator* 24 (March and April 1860). Reprinted in Hull, David L. *Darwin and his Critics: The Reception of Darwin's Theory of Evolution by the Scientific Community.* Cambridge: Harvard University Press, 1973, pp. 334-35.

Sloan, P. R. "The Sense of Sublimity: Darwin on Nature and Divinity." *Osiris* 16 (2001): 251-269.

Turner, F. M. "The Victorian Conflict between Science and Religion: A Professional Dimension." *Isis,* 69, no. 3 (1978): 356-376.

Young, R. M. *Darwin's Metaphor: Nature's Place in Victorian Culture.* Cambridge: Cambridge University Press, 1985.